3D Modeling of Buildings

Series Editor
Anne Ruas

3D Modeling of Buildings

Outstanding Sites

Raphaële Héno
Laure Chandelier

iSTE

WILEY

First published 2014 in Great Britain and the United States by ISTE Ltd and John Wiley & Sons, Inc.

ISTE Ltd
27-37 St George's Road
London SW19 4EU
UK

www.iste.co.uk

John Wiley & Sons, Inc.
111 River Street
Hoboken, NJ 07030
USA

www.wiley.com

Library of Congress Control Number: 2014941560

British Library Cataloguing-in-Publication Data
A CIP record for this book is available from the British Library
ISBN 978-1-84821-536-8

Printed and bound in Great Britain by CPI Group (UK) Ltd., Croydon, Surrey CR0 4YY

Contents

Acknowledgements

First of all, we thank Anne Ruas, Director of Research at IFSTTAR, who suggested that we start out on this adventure. Without her invitation, the work carried out throughout the last few years could not have been capitalized upon in this book.

This book also gives us the opportunity to pay a special tribute to Yves Egels, geographical engineer who throughout his career at IGN, has tirelessly contributed to the development of photogrammetry, in general, and more specifically to the use of this technique for the documentation of heritage. An essential point of reference in photogrammetry, we owe our expertize in the field of architectural surveying to him.

Without the expert advice of our proofreaders, Yves Egels, Patrice Bueso and Jean-Charles Pruvost, this book as it is today would not have been possible and, therefore, we are grateful for the time spent and the constructive criticism given.

Many people, teachers from ENSG and instructors from IGN, have worked with us on the architectural projects which are the cornerstones of this book. We thank them for their commitment and enthusiasm.

It is also important to thank the project sponsors who created the projects and worked with us in carrying them out. A special mention should go to Marylène Barret for Yemen and El Mustapha Mouaddib for Amiens Cathedral.

Finally, although it is not possible to name them all, we thank all the students of ENSG and elsewhere who worked on the projects addressed in this book.

Introduction

Traditional topographic databases, usually created from aerial or satellite imagery, provide a simplified three-dimensional (3D) model of our urban environment which can be used for a number of applications, such as town planning, risk prevention and mobility management. However, in the case of representing and analyzing outstanding sites like monuments, works of art or archeological sites, these databases are insufficient and other ways of acquiring and processing data need to be employed. This book presents a state-of-the-art of the methods specifically adapted to outstanding sites and the research currently being carried out in this area. The methods addressed in this text range from lasergrammetry to current dense correlation techniques using images as well as traditional photogrammetry.

These methods allow for the surveying of outstanding sites, permitting the restitution of the structures' form and appearance with a precision and a level of detail agreed upon in advance in a set of building specifications. They are due to takeover from techniques traditionally used for the creation of topographic databases, in the case of specific orders, such as the description of the relief of the façade and the complete 3D survey (interior and exterior) of a monument. In this text, digitization mainly involves 3D geometric rendering, and does not include the inventory, analysis of the construction materials, structural analysis of the building or its history.

First, it is necessary to define what is meant by the term “outstanding site”. These are buildings, collections of buildings, monuments or works of art which attract a level of interest to such an extent that any traditional 3D products which currently exist are not able to reproduce the vividness of the original. The size of the sites dealt with in this text is measured from ground level, including structures ranging from a small cellar to an entire district, as well as castles, churches, bridges and towers. Therefore, it is only static objects which will be dealt with in this text.

The aim is to give an overview of the range of current 3D surveying methods. It is neither a theoretical lecture, nor a collection of procedural manuals. The reader will find no detailed algorithms, but instead will find information on all working stages to create a 3D survey of an outstanding site. Advice and recommendations are also given, which the authors compiled paying close attention to produce a result which is as reliable, faithful and precise as the current methods will allow and should, therefore, be used simply as guidelines.

Some building specifications focus more on visual appearance than on overall accuracy. In this case, the demands on the data acquisition phase will be relaxed, while the post-processing cosmetic phases will be lengthened. Therefore, this text is relevant to anyone interested in high resolution 3D surveying, whether they are project managers, surveyors, heritage conservationists, town planners, computer graphics designers, or those who wish to learn about the technical methods available, possible methodologies and the steps to be followed in carrying out a project of this kind.

The text does not aim to provide an exhaustive list of methods or references but instead presents the views of the authors, which is manifestly influenced by their training and profession as photogrammetry engineers.

1

Specific Requirements for the 3D Digitization of Outstanding Sites

1.1. The current offer for high-resolution 3D data

In France, the IGN (*Institut national de l'information géographique et forestière* or the National Institute for Geographical and Forestry Information) produces and maintains the RGE[1] (*Référentiel à Grande Echelle* or Large-scale reference) over the entirety of the territory, containing four levels of information: a high-resolution orthoimage, an "addresses" database, a georeferenced cadastral map and a 3D topographic database of metric precision containing a description of elements of the landscape, classed according to type. Work is underway on a new-generation RGE, which is more precise and gives better resolution and detail, largely due to the cooperation of a number of stakeholders who collect and manage geographical information throughout the territory.

The IGN is also implementing a semi-automatic production process using high-resolution aerial images (pixel size below 20 cm), which are highly redundant, (significant overlap between images) leading to the production of vector

1 http://professionnels.ign.fr/rge.

databases in line with the LOD2 (*level of detail*) specifications of the CityGML[2] standards. It has been implemented in some large French cities, such as Paris, Nantes, Rennes, Besançon and Marseille. It shows the various districts, roof structures, vegetation and generalized street furniture. This product, Bati3D® [MAI 04], not only contains a survey of the buildings' forms (or groups of buildings if they are connected), but also their appearance, since the volume reproduced via automated photogrammetric processes is textured by aerial images. It is possible to improve the rendering and the precision of the façades and street furniture by adding data acquired in the field by vehicles equipped with stereoscopic cameras, lasers scanners and direct georeferencing devices (GNSS[3] antenna, inertial navigational system or odometer). These devices are rarely self-sufficient in urban environments, but are essential for initiating the georeferencing of images through photogrammetry.

Many local authorities (well-known examples include Monaco[4], Geneva[5] and the urban community of Lyon whose model is now available to be consulted on-line) have financed digital 3D models of their territory, believing that the traditional products available were not able to meet their requirements. In general, specific teams manage the production of Digital Elevation Models (DEMs), orthoimages and vector models representing buildings, street furniture and vegetation, with the use of aerial images and adapted lidar technology. Large internet-based companies are also working on 3D representations of towns, offering navigation

2 http://www.citygml.org/.

3 Global Navigation Satellite System.

4 http://www.gouv.mc/Action-Gouvernementale/La-Qualite-de-Vie/La-Prospective/Actualites/La-maquette-3D-de-Monaco-presentee-par-la-DPUM-a-Imagina.

5 http://ge.ch/sitg/cartes/3d.

services in the most realistic reproductions of our environment. An example of this is Google, which offers detailed reproductions of entire districts. A number of different methods are used to achieve this from traditional photogrammetric stereoplotting, through to pure infographics and the production from multi-images available to every Internet user[6]. This range of processes gives rise to questions regarding traceability of data and the amount of confidence which can be invested in it.

The quality of aerial images and the ever more automated computerized tools allow for the large-scale production of urban databases with increasingly detailed resolutions. However, the majority of these databases solely aim to show towns from an external view-point, without looking into the buildings or going underground.

On the other hand, the architect's digital model, or Building Information Modeling (BIM), is exhaustive, showing the building or the district from all angles: internal, external, underground and from above. Either this precedes the construction phase and gives an architect's view, or it is used in urban restoration projects (modernizing old buildings), and it meets the requirement of providing a model "as-built".

The methods of digitization discussed in this text provide a description of a particular building, which is more applicable than that given by other products available in terms of better spatial resolution or greater comprehensiveness. The text refrains from examining computer-aided design (CAD) processes, which concentrate more on the image than the accuracy of the survey. Here, the focus will be on geometric methods which, while keeping up-to-date, are able to remain objective and render a building as it really is and not as it may be imagined.

6 http://www.google.com/earth/learn/3dbuildings.html#.

1.2. Statement of requirements

The digitization methods which are addressed in this text involve the following stages [BAR 11]:

– presentation of requirements;

– choice of techniques to be applied;

– data acquisition in the field;

– data processing;

– quality control;

– layout and delivery.

1.2.1. *Potentials*

Drawing up the statement of requirements for digitization, prior to the production of building specifications for the contractor, is a challenging exercise. A comprehensive, precise and well geolocalized digitization of an outstanding site may be requested in the case of an inventory, a restoration or an archeological dig, or with the aim to make it accessible to everyone everywhere with a digital model that can be consulted online (virtual tourism). Requirements regarding content, resolution, precision or up-to-dateness obviously vary depending on the context. However, the statement of requirements is often initially summarized in an idealistic manner: the 3D model of everything at the highest resolution. Budgetary constraints tend to bring project sponsors back down to earth. The reasons motivating digitization, in terms of potentials, should, therefore, be highlighted. The following examples can be given:

– Understanding the monument:

- establish a model of movements;

- recover the history of its construction;

- analyze its distortions.

– Providing a collection of measurements:

- count openings;

- measure the surface of the roofs;

- display different sections of the structure;

- study the stability of the monument.

– Locate the monument in time and space:

- superpose elevation to previous or future surveys;

- create an evolving 3D model;

- visualize the site and its positioning within its environment;

- view installations and existing equipment;

- show future projects or constructions.

– Communicate, broadcast and exchange:

- broadcast the model online;

- consult the model online;

- be interoperable with a cadastral plan or other traditional geographic information system (GIS) layers;

- provide a 3D immersion.

– Administrate:

- take decisions on strategic projects;

- re-develop the building to enable visits from the public;

- visualize and present projects during the planning permission applications;

- allow tracking of town-planning regulations.

The conservation survey is essentially an exception to the building specifications. It can be used as a memory in case of deterioration, or even total or partial destruction, which can be due to a range of causes: “war, significant restructuring, but also disrepair and lack of maintenance” [SAI 92]. However, without being able to predict how the survey will be used in this case, it is difficult to establish a comprehensive set of building specifications.

1.2.2. *Conversion into building specifications*

These potentials should firstly be converted into product specifications to be delivered (nature, coordinate reference system, accuracy of georeferencing, spatial resolution, coverage, lead times, etc). Depending on the formulation of the order, the products to be delivered can be made available in a number of different forms, as shown by the examples below:

– point cloud showing the main structural lines of the building;

– point cloud to enable user to count mason’s marks;

– point cloud with a spatial resolution of 1 cm, with permitted noise of 2 mm;

– 3D line drawing showing the major structural lines of the building;

– 3D line drawing showing the major structural lines of the building and the stone façade;

– triangulated and textured 3D line drawing;

– contour lines giving volume to a dome;

– façade elevations;

– sections, etc.

As well as the aspects to survey, the building specifications must state the precision of the associated geolocalization. The internal precision of the survey must be coherent with the survey's resolution because, if this is not the case, significant geometric gaps could be interpreted as characteristics of the structure. Demands may often be relaxed on the absolute precision of localization; positioning in a given space is useful in terms of situating the structure in its environment, or to understand the principle axes in relation to the cardinal points for instance. The precision necessary can, therefore, be decimetric or even metric.

If the project sponsor makes the added requirement of communicating, broadcasting and exchanging information on the 3D model produced, then post-processing of the 3D point cloud will be necessary to lighten it, structure it and add semantic information. If the order is more focused on the interoperability of the survey, especially with the requirement of superposing pre-existing data onto the new model, special care needs to be taken regarding the absolute georeferencing. It is, unfortunately, impossible to establish simple links between orders and the product due to the combination of many different parameters that must be taken into account, such as the budget, schedules and preparation, thus making each project unique.

1.2.3. *Technical survey specifications*

The specifications of the product requested will be converted into technical survey specifications by the contractor: laser and/or image acquisition (resolution, provisional number of scanners or photographic equipment), georeferencing equipment (how, how many and where?), material and software necessary for post-processing [BRY 09].

Evidently, the optimal spatial resolution of images depends on the requirements; to produce a model of a building's major structural lines, a centimetric or even decimetric resolution is sufficient. To be able to model details, a better resolution is necessary (e.g. 1 cm to show the stone façade, 2 mm to represent the mason's marks). In the era of analog photography, it was advocated that a plan of 1/100 was suitable for a general building survey, a plan of 1/50 for a more precise survey with stone façade detail and a plan of 1/20 for a very detailed survey including stonework, sculptures and inscriptions. Regarding the choice for the resolution of the point clouds according to the output scale, [COL 10] advises resolutions of 2, 4, 15 and 25 mm, respectively for plans of 1 /10, 1/20, 1/50 and 1/100.

These values are given only as an indication, and it is strongly advised to check that the acquisition resolution chosen is able to capture all the elements required for each structure.

1.3. Existing surveying methods

1.3.1. *Existing acquisition methods*

Digitizing an outstanding site means faithfully recording its form and appearance, possibly in a coordinate reference system. The data produced from this are delivered in a format, which has been decided upon in advance. According to requirements and means, the methods used may vary. The simplest method, but not the quickest nor the most reliable, is the direct manual sketch as carried out by architects and archeologists. This is done by attaching copy paper to the areas, which need surveying or by using common measurement equipment, such as a distance meter or a plumb line. It is simply unrealistic to imagine the production of a high-resolution 3D model with this method. Other delivered models can be produced by extrapolating existing plans [REM 09].

Indeed, surveyors are very skilled at quickly digitizing a building with a theodolite, but this process inevitably involves a significant simplification of forms, determined by individual interpretations and choices. A survey with a laser scanner automates this process to some extent, since the equipment takes a systematic scan of the area according to a preset angular interval, which may be very precise, while the theodolite is only able to target isolated points. The initial result of a digitization from a laser scanner is a very dense 3D point cloud showing the surface of the building. The final model is produced by placing this next to neighboring point clouds situated in a unique reference system by appropriate methods. The result is rarely the point cloud itself, but plans, cross-sections and elevations that can be derived from semi-automated methods, or a lighten, triangulated and textured model. Photogrammetric methods – those based on stereoscopic images – can be used to produce 3D images, unrolled images and orthoimages and colored 3D point clouds.

For the majority of sites, and especially for outstanding sites, a combination of these methods would be chosen in order to optimize the value for money and to deal with the diversity of situations that one is bound to encounter [REM 09]. This text will examine the respective qualities of these techniques in terms of rapidity, accuracy, simplicity of implementation and processing as well as reliability.

1.3.2. *Example surveys*

The table below shows examples of surveys, which highlight the diversity of requirements and the technical solutions possible.

Site	Date	Requirement	Techniques used	Products delivered
Temples of Abu Simbel (Egypt) by UNESCO – IGN France [BON 58]	1955	Detailed survey of the two rupestrian temples of Abu Simbel subject to its relocation before the construction of the Aswan dam	"Traditional" photogrammetry Stereoscopic image surveying (with photographic equipment designed for aerial photogrammetry) to provide a representation with contour lines for the various elements Pegging to a unique coordinate system	Plans and elevations of the various surfaces which make up the structure Precise descriptions of surfaces with contour lines (scales requested: from 1/10 to 1/150)
Lascaux Caves by IGN France [HOU 80]	1966	Real scale reconstruction of the cave	"Traditional" photogrammetry (stereoscopic image surveying and graphic restitution)	Longitudinal and transversal cross-sections every 25 cm on a scale of 1 in the Salle des Taureaux; texturing with color photos on a scale of 1 to produce an exact copy, as displayed at the Grand Palais in 1980
Tower of Pisa by IGN France [IGN 92]	1992	3D model of tower to support further analysis	Precision topometry to locate a number of targets Photogrammetric image surveying	3D photogrammetric coordinates of all targets Cross-section plans of the tower

Saint-Guilhem le Désert[7] by MAP[8], UMR 694	2007	3D model of cloister and all dispersed elements	Long-range 3D laser survey Acquisition of images for visual appearance and conservation state of materials	Virtual anastylosis of surrounding fragments Space/time information system on an architectural level
Karnak by ATM3D – ENSG[9] [CHA 09]	2008	Photographic reviews of the 134 columns in the Hypostyle hall of Karnak temple	3D laser survey of columns Acquisition of stereoscopic images with the use of poles (32 images per column) Pegging to a unique coordinate system	Unrolled image of each column
Cluny by ENSAM [PÈR 10]	2009	Digital model of the biggest 12th century Romanesque abbey church	3D laser survey (time-of-flight for remnants of the structure and triangulation for the stonework)	Reverse-engineering of remaining elements using 3D point clouds
Amiens Cathedral by ENSG – Laboratoire MIS [HÉN 11]	2010 – 2013	Complete 3D model to serve as a database for historians, and digital archive and virtual visit tool for conservationists	3D laser survey (resolution adapted to the complexity of the object – from several mm's to 2 cm) Very high-resolution aerial and ground image survey to produce automated 3D point cloud	Network of geodetic points documented Georeferenced 3D point clouds in the French legal system Georeferenced panoramic images in the French legal system

Table 1.1. *Examples of surveys*

7 http://www.map.archi.fr/3D-monuments/telechargements/sgdesert_doc-ecrit.pdf.

8 http://www.map.archi.fr/

9 Ecole nationale des sciences géographiques, http://www.ensg.eu (National School of Geographical Sciences).

1.3.3. *Survey control data*

The current demands for the interoperability of surveyed 3D data usually require them to be georeferenced in a coordinate reference system (e.g. RGF93 in France). To do so, surveyors use or implement a network of geodetic points which is easy to densify into ground control points for the specific georeferencing requirements of laser data and images. Such operations are familiar to surveyors and require their knowledge and equipment (GNSS receiver, theodolite, and topometric calculation devices). For simple sites, it is sometimes possible to simply provide orientations (vertical measurement ascertained with a plumb line or level, and north confirmed by a compass) and scaling provided by a distance on the structure. In a general way, this text will refer to a global coordinate system, that is to say any coordinate system which enables comprehensive georeferencing of everything in one single coordinate system, whether it refers to the structure itself or to a reference system.

1.4. From building specifications to realization

1.4.1. *Reconnaissance stage*

Reconnaissance of a site starts with a visit of the area to be surveyed as detailed in the building specifications. It involves the precise definition of elements to digitize, including the resolution and accuracy of georeferencing demanded. It is useful for the project sponsor to be present in case of any ambiguities emerging from the building specifications.

For the team tasked with the data acquisition, this step will focus firstly on the positioning of the various measurement instruments (laser scanner, tripod, theodolite, etc.) To save time and simplify the acquisition process, attempts should be made to minimize the number of laser stations needed. Photographic stations shall be comprehensively defined (at this stage of the process, the exact number of cameras is not yet established) according to the photographic material available (different photographic material will mean that the distance from the object of interest will vary). Evidently, access authorizations to all areas directly involved in the survey process must have been negotiated before this reconnaissance phase, as well as the possibility of diverting the general public for a short period in order to have undisturbed access to part of the structure. Indeed, it is common to be confronted by one or more of the following on-site problems: obstacles or restricted access to certain areas, continued public access, no access to certain details, short-lived site access, etc. If necessary, requests should also have been made for authorizations to attach targets or to make holes in which to place geodetic reference points.

When the positioning of the laser scanners and photographic equipment is decided upon, the topometric stations will be optimized (tripod and theodolite) from where ground control points will be measured which can be used to georeference laser data and images. These points may be of differing types, but will always be chosen so as to be visible in a laser point cloud or on images, depending on what needs to be georeferenced:

– natural points of detail: characteristic elements of the site, stonework, paintwork detail;

– artificial targets: material (wood, paper, plastic and ceramic) with geometrically characteristic symbols (cross, point and circle) or shapes (sphere), applied to the site through collage or magnetization;

– geodetic points: long-lasting elements (bronze work and nails) placed on the site to facilitate a return visit to the site, potentially used as a control point.

The use of one control point over another is decided by:

– the site's limits (whether it is possible to implement a given point);

– ease of implementation and traceability;

– precision of localization caused by the detail: the precision of targets is generally high, whereas a piece of natural detail is more variable, depending on its geometric properties, and is sometimes not as precise;

– the need for a return visit;

– the requirement of having a network of registration points with reliable geometric properties.

For a complex site, it is important to be mindful of the diversity of the registration points which can be used to provide a practical, efficient and flexible solution. Finally, regardless of the nature of the registration points used, it is necessary to have enough to enable the registration or georeferencing of the data acquired. Further details on registration points to provide will be given in Chapters 2 and 3.

Following the reconnaissance phase, a site plan should be created, including the positioning of the various laser scanner stations, cameras, theodolites and points to be measured, as well as an adapted numbering system. In general, the numbering system should unambiguously distinguish between the nature of the point (control point or station with an instrument) and its position (interior or exterior, cardinal points and upper/lower).

Ideally, the site should be treated as a whole, even if the order is only relevant to part of the site. This is sometimes the case for outstanding sites where its size and complexity mean that the whole project cannot be carried out in one go, due to budget limitations, accessibility problems or bad weather. It may, therefore, be necessary to return several times to the same site to carry out data acquisition as and when it is possible to do so. The building specifications should take this into consideration and clearly specify elements which need to be implemented, enabling coherent and efficient time management. This can be achieved largely by defining ground control points or elements which are sufficiently long-lasting, stable and precise.

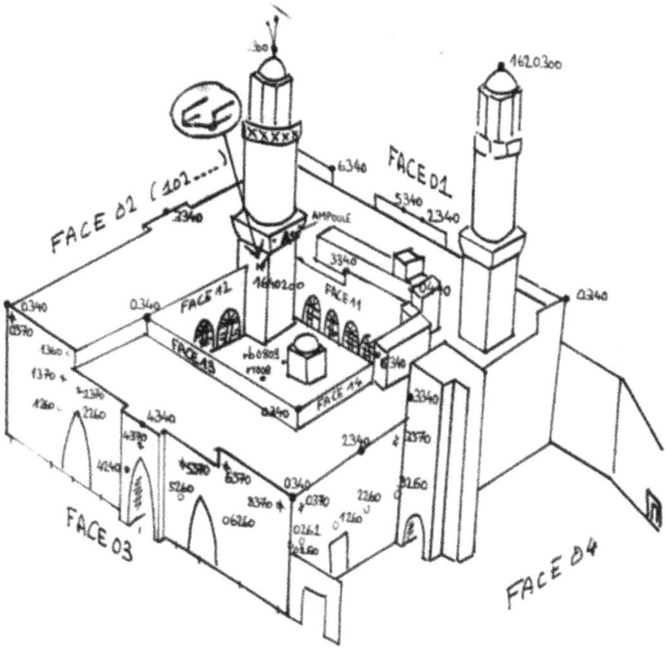

Figure 1.1. *Diagram of the Great Mosque of Sana'a with proposed numbering system*

1.4.2. *Acquisition*

When the site plan has been fully completed, the acquisition stage can start. Providing the reconnaissance stage was sufficiently comprehensive, it should *simply* be a case of following the plan and managing any outside constraints that may arise.

Operations are carried out as stated, depending on:

– the number of workers;

– the number of instruments available and their technical characteristics;

– whether or not it is possible for several people to work in the same area without being inconvenienced;

– the weather forecast for any acquisition being carried out outside;

– possible presence of the general public in some areas of the site;

– lighting (mainly for photographic equipment).

1.4.3. *Processing of data*

Where possible, data processing should start at the site in order to verify whether anything needs to be redone. It is necessary to verify the data's integrity (is it possible to open and read the files?), comprehensiveness (is the entire area of interest covered?) and conformity with requirements given (have the resolution and stereoscopic overlap specifications been respected?). For safety, this initial data from the measurement instruments should be duplicated and archived. The images may be named by area and by camera. Topometric adjustment calculations should be started as soon as possible so that measures can be retaken if there is a problem with the originals.

Back in the office, laser and image data must be georeferenced in the project's coordinate reference system. If possible, georeferencing operations can be carried out immediately after data acquisition, possibly while still on site, since it can be easier to detect any shortcomings in the data collected [BOR 10].

Subsequently, depending on the products required by the building specifications, all or some of the processing steps will be carried out from the table below. For each stage, an estimated time is given, the order of magnitude corresponds to a medium-sized outstanding site, such as a church or small castle.

Stage	*Processing time*	*Of which is automated processing*
Field data acquisition		
Reconnaissance	1 day	
Acquisition (topometry, GNSS, images, laser)	6 days	
Preliminary processing: verifying integrity and quality of data, specifications have been respected, points have been properly named	1 day	
Data processing: georeferencing of data		
Topometric calculation of ground control point network	1 day	
Georeferencing of images	*2 days*	*Calculation of tie points: 1 day*
Georeferencing of laser stations	*1 day*	
Data processing: use of data		
Calculation of 3D point clouds using images	*4 days*	*Dense correlation: 3 days*
Pre-processing of point clouds	*1 day*	*Sampling, denoising, etc.: 1 day*
Manual stereoplotting	*Varies depending on specifications: survey of major lines, stonework and sculptures => 30 days*	

Line drawings and 3D reconstruction by digitizing point cloud	*Varies depending on specifications: survey of major lines, stonework and sculptures => 30 days*	
Meshing or semi-automatic and texture reconstruction	*Varies depending on specifications => 6 days*	*Manual operations expected, time varies depending on software tools used.*
Generating derived image products (orthoimages, etc.)	*2 days*	*Orthorectification, radiometric equalization: 1 day*
Layout and delivery		
Orientation, delivery, archiving, etc.	1 day	

Table 1.2. *Example of workload allocation for each stage of the process (in italics, stage carried out depending on the acquisition technique used and products requested)*

1.5. Bibliography

[BAR 11] BARBER D., MILLS J., *et al.*, *3D Laser Scanning for Heritage (second edition) – Advice and Guidance to Users on Laser Scanning in Archaeology and Architecture*, English Heritage, 2011.

[BON 58] BONNEVAL H., "Utilisation de la photogrammétrie aux levés architecturaux. Application au temple d'Abou-Simbel (Haute-Egypte)", *Communication de l'Institut Géographique National au Congrès des Sociétés Savantes,* Paris, 1958.

[BOR 10] BOREL L., CABARROU M., DUBOURG S., *et al.*, "D'X,Y à X,Y,Z, de nouveaux outils pour l'étude architecturale et archéologique. Restitution 3D, lasergrammétrie et photogrammétrie: le cas de la citerne el-Nabih à Alexandrie", *Actes du colloque Virtual Retrospect 2009*, Pessac, 18–20 November 2009, pp. 151–159, *Archeovision*, vol. 4, Editions Ausonius, Bordeaux, 2010.

[BRY 09] BRYAN P., BLAKE B., BEDFORD J., *et al.*, *Metric Survey Specifications for Cultural Heritage,* English Heritage, 2009.

[CHA 09] CHANDELIER L., CHAZALY B., EGELS Y., *et al.*, "Numérisation 3D et déroulé photographique des 134 colonnes de la Grande Salle Hypostyle de Karnak", *Revue XYZ*, vol. 120, pp. 33–39, 2009.

[COL 10] COLOMBO L., MARANA B., "Terrestrial laser scanning", *GIM International*, vol. 24, December 2010. Available at http://www.gim-international.com/issues/articles/id1629-Terrestrial_Laser_Scanning.html.

[HÉN 11] HÉNO R., CHANDELIER L., CLAUZEL L., *et al.*, "Complementarity of terrestrial laser scanning and DSM from aerial photography: the example of the cathedral of Amiens", *Proceedings of the 23rd CIPA Symposium,* Prague, Czech Republic, 2011.

[HOU 80] HOURS M. (sous la direction de), *La vie mystérieuse des chefs-d'œuvre: la science au service de l'art,* Éditions de la Réunion des musées nationaux, p. 334, 1980.

[IGN 92] IGN, GEOFIL, *Magazine de l'Institut géographique national*, 1992.

[MAI 04] MAILLET G., FLAMANC D., "Une approche opérationnelle pour la production de modèles 3D urbains", *Revue Française de Photogrammétrie et Télédétection*, vol. 176, pp. 14–20, 2004.

[PER 10] PERE C., LANDRIEU J., ROLLIER-HANSELMANN J., *et al.*, "Reconstitution virtuelle de l'église abbatiale Cluny III: des fouilles archéologiques aux algorithmes de l'imagerie", *Actes du colloque of Virtual Retrospect 2009*, Pessac, 18–20 November 2009, pp. 151–159, *Archeovision*, vol. 4, Editions Ausonius, Bordeaux, 2010.

[REM 09] REMONDINO F., EL-HAKIM S., GIRARDI S., *et al.*, "3D virtual reconstruction and visualization of complex architectures – The "3D-ARCH" project", *Proceedings of the ISPRS Working Group V/4, Workshop 3D-ARCH "3D Virtual Reconstruction and Visualization of Complex Architectures"*, Trento, Italy 2009.

[SAI 92] SAINT-AUBIN J.-P., *Le relevé et la représentation de l'architecture,* Documents et méthodes, vol. 2, 1992.

2

3D Digitization Using Images

2.1. History in France

Although it seems unnecessary at this juncture to restate the history of photogrammetry (we refer the reader to [LAU 99, ROU 36] or [VIN 13]), it is interesting to highlight the fact that the first projects which used this technique were on outstanding sites. An early example of this comes from Aymée Laussedat who, from 1850, produced a plan of the buildings opposite his office by using a camera lucida. In Germany, at the end of the 19th Century, Meydenbauer institutionalized photogrammetry for the architectural surveying of monuments [ALB 01].

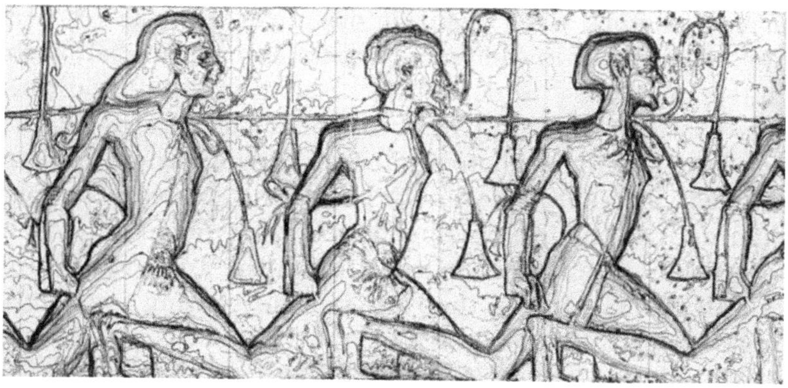

Figure 2.1. *Abu Simbel, Egypt, image XVI [IGN 67]*

In France, one of the pioneers of photogrammetry in architectural surveying was Henri Deneux [DEN 30], Chief Architect of Reims Cathedral, who developed his own methodology to obtain graphical reconstructions of monuments. Soon after its creation in 1940, the *Institut géographique national* (IGN) (The National Geographic Institute) was given the task of examining the potential of using photogrammetry to document architectural heritage. The founder of the French photogrammetry industry, Georges Poivilliers [VIN 13], carried out this study. He conducted the first tests in the winter of 1944 in the Sainte Chapelle in Paris, which were presented at the International Photogrammetry Congress in the Hague [IGN 48]. Photogrammetry quickly showed its potential in the field of architectural documentation, enabling the separation of operations in the field, the creation of plans and the production of varied documents as well as providing a completely objective way of representing the form of a monument at a given date [ICO 72]. Up until the start of the 1990s, the IGN carried out major documentation projects of architectural heritage using photogrammetry, both in France and around the world. Some emblematic projects include the great French cathedrals (Strasbourg, Saint-Denis, chevet of Notre Dame de Paris, Amiens, Rouen), projects connected to restoration work (Pantheon, Beauvais Cathedral, Lascaux cave), and work with scientific objectives such as research into stone calculi in Moissac. Projects abroad, funded by the United Nations Educational, Scientific and Cultural Organization (UNESCO) or the Ministry of Foreign Affairs, raised the profile of photogrammetry further. These projects included the Acropolis in Athens, the Nubian temples (Figure 2.1), the Borobudur temple, the Royal Palace in Kathmandu, the Pagan pagodas, up to the Tower of Pisa in 1992 [IGN 92]. Most of the restitutions resulting from these projects can still be consulted today at the IGN's map library in Saint-Mandé.

Over the same period, a special service was set up by the Ministry for Culture to put together an exhaustive and systematic national inventory of tangible heritage, started in the 1960s by André Malraux. It was operated by a department based in Paris specialized in stereoplotting and by regional teams mainly made up of volunteers to carry out work in the field.

From the early 1990s, surveys of outstanding sites with photogrammetric methods were only being carried out by private surveying agencies and university teams, which had the professional equipment necessary and honed skills in the photogrammetry field. Research laboratories which carried out surveys were usually made up of experts connected to professions in the surveying field [GRU 94]. The production method remained greatly unchanged: line cross-sections and elevations from manual stereoplotting, sometimes accompanied by orthoimages until the start of the 2000s when the mass arrival of laser scanners caused an overhaul of operating methods, which turned the traditional surveying field on its head and led to truly 3D renderings.

2.2. Equipment

2.2.1. *Equipment through the years*

Initially, photogrammetry required the use of metric cameras whose optical and mechanical characteristics were calibrated with great levels of accuracy in the laboratory (see Figure 2.2).

Images were taken so as to create stereopairs, with an overlap of at least 50% of the area of interest, with enough so as to fully cover the monument, according to the requirements (survey scale requested, distance from monument, hidden parts and blind spots).

Figure 2.2. *Examples of metric cameras [ICO 72]*

Topometric measurements, recorded simultaneously in the field, enable the establishment of a ground control point network for photogrammetric use of the images. The use of these images for photogrammetric reconstruction in the shape of elevations, plans, cross-sections and contour lines would be done with so-called "universal" photogrammetry equipment, because it allowed the processing of not only aerial images, but also other types of images [BON 58]. This equipment, based on the principles of mechanics and optics, allowed the user to view the photographed scene with the original relief. It also allowed him to draw in the notable elements with an illuminated measuring mark linked to a pencil, with which the user could trace the contours of the original forms.

Analog photogrammetry equipment was replaced in the 1970s by what is known as analytical photogrammetry equipment, enhanced by the use of a computer enabling the implementation of an image's formula, meaning the fundamental rules modeling the relation between a point on a photographed scene and its image on the photograph. This was still cumbersome (see Figure 2.3), since plate holders and devices to move into the stereoscopic model were required (often a handwheel for movement over the reference plane and a footdisk to move in the direction of depth). However, this equipment reduced the image surveying specifications, since the implemented model enabled more data processing.

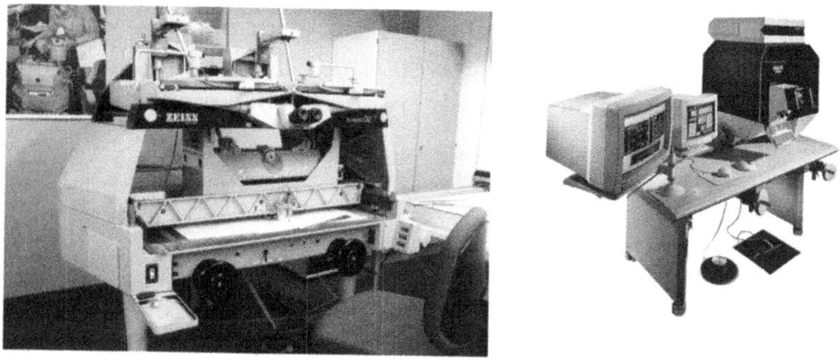

a) b)

Figure 2.3. *Examples of photogrammetric stereoplotting equipment. a) Analog stereoplotter (Planimat, Zeiss) b) Analytical stereoplotter (SD2000, Leica)*

2.2.2. *Modernization of equipment*

It was not until the 1990s that the first digital photogrammetric equipment further pushed back the limits of photogrammetric surveying. Digital photographic equipment quickly won over photogrammetrists who saw the possibility of using these high-quality images to add automatic measurement algorithms into their production processes, which had already proved successful for scanned analog photographs [MAÎ 92]. Today, a digital photogrammetric workstation is a standard computer, with all requirements regarding fluid manipulation of images possible on the computer screen, or even with a stereoscopic display, as these now make up a part of the normal configurations of mass-market computers (except for graphics cards on occasion).

The recent trend toward open-source software, which is usually automated, has made photogrammetry available to the masses, and no longer the privilege of a select number of specialists. Today, it is exceeding the circle of traditional cartographers to invest, notably in professions linked to heritage. Indeed, it is sometimes not even necessary to have a complete workstation, given that 3D reconstruction is automated and has partially replaced 3D manual plotting (see section 2.5).

2.2.3. *Cameras*

Current photogrammetric software are not only able to manage all of a camera's geometric characteristics (focal length and law of optical distortion) but also, in most cases, to determine them by using specific images acquired on a calibration site or even on any surface (auto-calibration, see section 2.4.2). Therefore, it is no longer necessary to use metric cameras, which have been calibrated with great accuracy in a laboratory. However, not all cameras guarantee the same image quality and differ in terms of their stability over time. The skills of the photographer should also be treated as a significant factor.

The choice of one piece of photographic equipment over another should be dictated above all by the requirement to capture greater or fewer details in the images (meaning the spatial resolution of images required, governed by the combination of distance from the object of interest and the focal distance of the photographic lens being used, see Figure 2.6). Other requirements, such as recording information in the infrared channel, controlling the camera remotely and authorizing a crash (especially for image surveying with a UAV), can restrict the range of possibilities further. It is possible to use bridge or compact cameras, but the fact that they are not fully stable and cannot be manually set are factors that could affect the quality of the photogrammetric product. The use of reflex cameras is advised, if possible with multiple lenses (see Figure 2.4) in order to adapt to all situations in the field (wide angle, such as a fisheye lens which, for instance, enables effective photography of narrow corridors, or a telephoto lens with which difficult-to-access details can be well captured, as well as "normal" lenses for more commonplace situations).

Figure 2.4. *Example of reflex photographic equipment with different lenses (15 mm, 24 mm, 35 mm, 50 mm and 100 mm)*

Analog cameras

Since digital cameras have demonstrated their prowess in many different ways (radiometric quality and flexibility of use), analog cameras are almost never used nowadays for photogrammetric surveying. However, it is possible to use images which come from the digitization of old photographs (Figure 2.5). The most difficult aspect in this case is to find the camera model as well as the geometric characteristics: focal distance, law of optical distortion and possibly fiducial markers, which would help us to define the precise link between the digital "image" coordinate system and the camera frame coordinate system.

Figure 2.5. *Scanned analog images*

2.2.4. *How do you measure with a camera?*

2.2.4.1. *Modeling the raw material: the image*

The image taken by a traditional camera (directly descended from Leonardo da Vinci's camera obscura) is modeled by the projection of the real world (in three dimensions) onto a two-dimensional surface (the camera image plane). Clearly, the geometric properties of this projection depend on the camera being used.

To simplify, the photographic lens of the camera, which is indeed composed of a number of smaller lenses, can be considered as one thin lens. In this case, optical laws dictate that any light ray passing through the center of the lens does not deviate. Therefore, we can schematically represent the creation of the photograph, or image, in the following way (Figure 2.6):

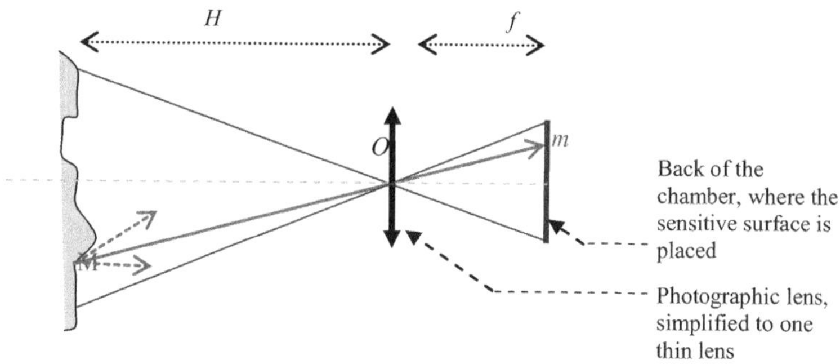

Figure 2.6. *Diagram of the creation of an image*

2.2.4.2. *Spatial resolution of images*

The dimensional sizes are:

– distance from the object, marked *H*;

– focal distance of the camera, marked *f* (distance from the optical center *O* of the lens to the surface where light rays from the infinite converge and imprint a sharp image on the back of the camera chamber);

– the dimensions of the image plane (where the sensitive surface is placed, film or charge-coupled device (CCD) or complementary metal oxide semiconductor (CMOS) sensor, which captures the image);

– the size of a pixel for a digital camera, marked *Tpx*.

The spatial resolution of an image (also called ground sample distance), henceforth referred to as GSD, denotes the size of the area covered on the photographed object by one pixel. It depends on *H*, *f* and *Tpx*, as in the following formula:

$$GSD = H \times \frac{Tpx}{f}$$

As stated in section 1.2.3, the resolution of an image is chosen depending on the building specifications: a resolution of several centimeters can be sufficient for the major structural lines of a building. To capture the stonework, a 1 cm resolution would be necessary. If even finer details are needed to be captured on the final product, an even better resolution would be chosen, at least locally, for the final image.

Evidently, for any given image, the spatial resolution is only an average value (see Figure 2.7, where one pixel on the ground corresponds to a greater surface than one pixel on the balustrade).

Figure 2.7. *Variable spatial resolution in the same image*

2.2.4.3. *Condition of collinearity: image formula*

In order to use images as measurement instruments, it is necessary to know the precise relation between a point object

M and its image *m*. This relation is known as the image formula, expressed by the mathematical function $m = \phi(\mathrm{M})$. Figure 2.6 shows that a point object *M* diffuses the incident light in all directions, and that one of the rays which are reflected enters the camera and imprints the image *m*. This is why the image formula is called, to a first approximation for images taken by a traditional camera, the condition of collinearity, since it shows that *M*, *S* and *m* are aligned [KAS 01].

It is also necessary to take account of some differences between reality and this model, in the case where their effect is higher than the accuracy expected of this method (the curvature of the earth and atmospheric refraction in aerial photogrammetry and optical distortion in all cases). In close-range photogrammetry (as opposed to aerial photogrammetry, which is farther from the object of study), optical distortion is the most important factor to be taken into account.

2.2.4.4. *Determining the parameters for the image formula: photogrammetric orientation*

The image formula, essential to any photogrammetric production, depends on the position (*S*) and the orientation (*R*) of the camera at the time the images are taken, as well as the geometric characteristics of the camera itself. These geometric characteristics (see Figure 2.8) are the principal points (principal point of auto collimation *(PPA)* and principal point of symmetry (*PPS*), the law of distortion (see Figure 2.9) and the principal distance (or the distance between the optical center and the image surface, different from the focal plane with close-range focusing). These should be defined, either beforehand by a calibration procedure in the laboratory or on a specific site known as a calibration site, or by using the images which will establish the photogrammetric survey themselves (henceforth known as autocalibration).

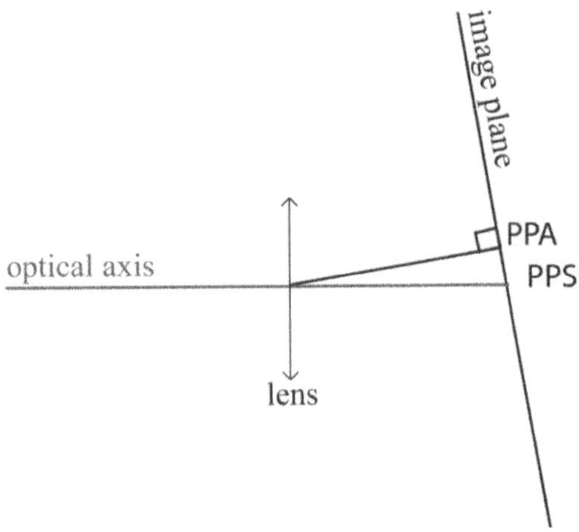

Figure 2.8. *Diagram of a photographic lens*

Figure 2.9. *The effect of optical distortion*

The external parameters of the camera (S and R) are either measured during the image surveying process by specific devices (global navigation satellite system (GNSS) antenna, inertial system, especially used during aerial image surveying) or determined by photogrammetric orientation (see section 2.4).

2.2.4.5. *Using images taken*

Once the images have been taken, it is possible to:

– reconstruct visible details in the images in 3D by manual capture;

– automatically produce 3D point clouds;

– orthorectify the images (possible to achieve with knowledge of the relief of the area of interest).

2.3. Image acquisition

2.3.1. *Characteristics of the images*

The principal qualities required for technical images of a photogrammetric site, in order to satisfy the need to be able to correctly view the object of interest in the image, are the following:

– correct exposure (no area which is over- or underexposed);

– sharpness of each point (in other words, large depth of field);

– realistic and consistent colors;

– as far as possible, fixed focusing per subset of images so that the geometric characteristics of the camera that affect the model can be precisely determined by photogrammetry.

The diagram below shows a camera (Figure 2.10), where the various components that influence the quality of the image are represented.

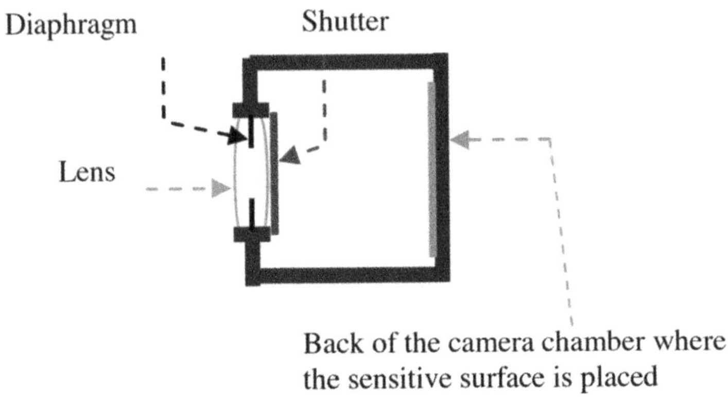

Figure 2.10. *Simplified diagram of a camera*

2.3.1.1. *Exposure*

When the camera has been chosen and for reflex models the lens has been selected, correct exposure is achieved by adjusting the lighting of the object of interest, exposure time (aperture time of the shutter for the light to enter into the camera), aperture of the diaphragm (which adjusts the amount of light that can enter the camera) and the sensitivity of the sensor (to emphasize the entering signal). The exposure is considered as correct when the histogram of colors is balanced (number of pixels per color, see Figure 2.11), and when there are no saturated colors. To help improve the realistic nature of the colors by postprocessing of the original images, a standard color chart can be included in the image.

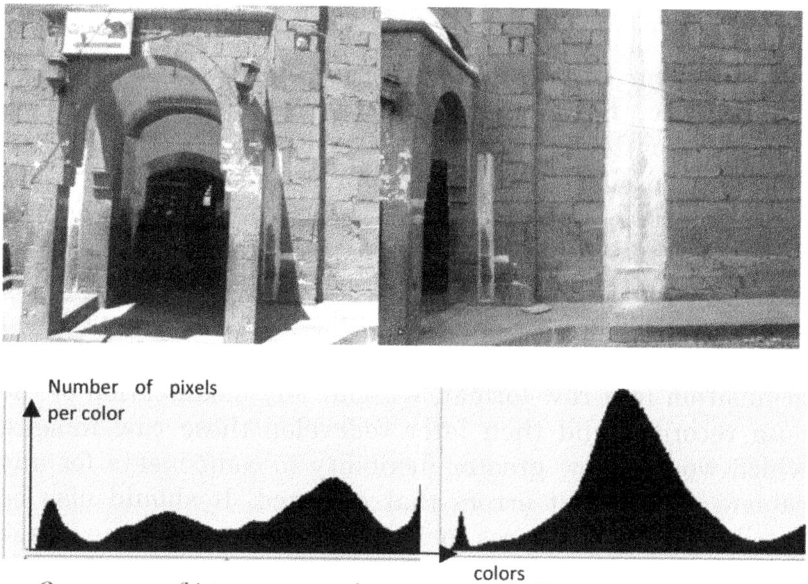

a. Overexposure (histogram cropped at the right) *b. Correct exposure*

Figure 2.11. *Different exposures on two images showing the south exterior wall of the Great Mosque of Sana'a (Yemen)*

2.3.1.2. *Sharpness*

The sharpness of an image is adjusted by the focus range. The depth of field, or the area in the scene occupied by sharp objects on the image, can be increased if [FER 04]:

– the diaphragm is relatively closed (high aperture number, around 9 or 11 depending on the camera; higher values can cause unfortunate diffraction effects);

– the focal length is short;

– the focusing distance is large.

To achieve a large depth of field, limiting the risk that certain areas of the image are blurred and therefore unsuitable for photogrammetric processing, it is preferable to work with the diaphragm in the closed position.

2.3.1.3. *Image format*

Finally, images should be recorded in a format which does not degrade the data, so that the automatic applications based on pattern recognition algorithms are the highest quality possible. Most compressed formats (Joint Photographic Experts Group (JPEG)[1] or Enhanced Compression Wavelet (ECW)[2]) can indeed reduce the volume of data to process, but the loss of data that this involves also leads to artifacts (visible smear effects on high compression JPEG formats). It is also possible to carry out this acquisition in a raw format, without any modification of the data recorded and then later redevelop these raw images, which would allow greater flexibility to compensate for any camera adjustment errors that occurred. It should also be noted that most cameras include metadata with some image formats which show the camera settings at the time an image was taken (known as Exchangeable image file format (EXIF)[3] data).

2.3.2. *Traditional stereoscopic image survey*

Every photogrammetric process requires each point of the photographed scene to be visible on at least two images. In other words, the images must overlap by at least 50%. Until the recent introduction of highly automated software tools, stereoscopic image surveying, carried out by procedures of manual stereoplotting or orthorectification, was performed by strips with small areas of overlap (sidelap of 10% to 20%). Production involving a visualization of stereoscopic pairs by an operator requires a "normal" stereoscopic configuration, whereby the images are at parallel axes, and perpendicular to the area of interest. Such image surveying specifications

---

1 http://www.jpeg.org/.

2 http://geospatial.intergraph.com/products/ecw/details.aspx.

3 http://www.cipa.jp/std/documents/e/DC-008-2012_E.pdf.

(sidelap of 20% and forward overlap of 60%; axes parallel to each other and perpendicular to the area of interest) are simple to implement from an aerial point of view for a low-relief landscape. Indeed, in this case, the trigger of the image surveying is aided by GNSS methods at a trajectory calculated according to the image surveying specifications and the relief of the ground. This is not the case for terrestrial image surveying, which is confronted by the front of structures which are extremely complex relative to the distance of the camera, as well as hidden areas, a lack of perspective, and an inaccessibility of some angles (the upper parts of façades especially). Whatever the context, it is essential to respect the constraint of having a B/H (relation between the stereoscopic base and the distance to the area of interest) of between 1/2 and 1/6 to ensure that the operator is comfortably able to take measurements and guarantee the correct level of accuracy regarding depth restitution (Figure 2.17) [WAL 94].

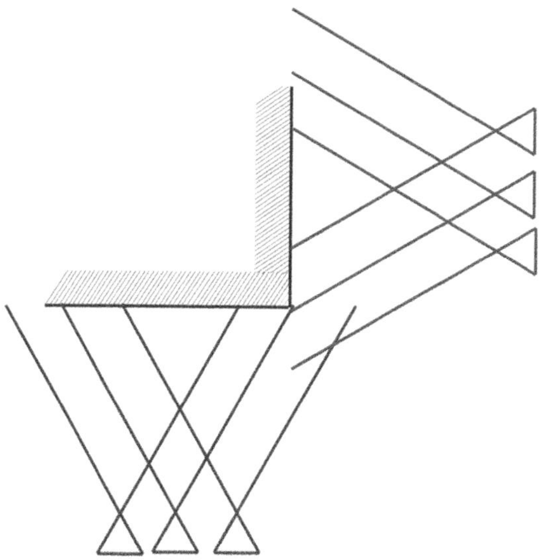

Figure 2.12. *Traditional image survey of areas of interest on the corner of a building (as seen from above)*

2.3.3. *Stereoscopic image surveying for the automatic production of 3D point clouds*

Most of the image surveying carried out nowadays is much more redundant than traditional image surveying. Indeed, the constraint has changed from requiring all points of the area of interest to be visible on at least two photos in normal configuration (for the requirements of manual stereoplotting by an operator), to requiring all points of the area of interest to be visible in at least three photos (for the reliability requirements of automatic restitution algorithms) [PIE 11].

For the use of new automatic software tools, it is advised that additional rules be introduced. These are listed below in the order of importance:

– *Increase overlaps to between 60% and 80% to ensure that the matching algorithms are successful*

Indeed, the new photogrammetric tools which, among others, enable the production of 3D point clouds are based on the automated matching of homologous details (see section 2.5). To guarantee the success of the automatic matching process, the viewpoints between the various images must not be too different (the overlap must therefore be greater, see Figure 2.13).

– *Do not hesitate to take convergent photos*

Indeed, by increasing the angle of intersection of homologous perspective rays, the accuracy of the 3D reconstruction can be enhanced. Moreover, these photos are essential for some objects which are truly 3D, as opposed to those at parallel axes that are used for large and relatively plane surfaces (Figure 2.14).

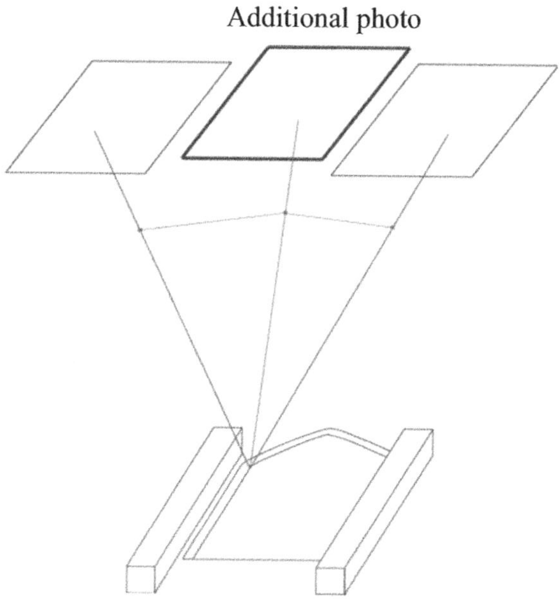

Figure 2.13. *Additional images to ensure successful matching algorithms*

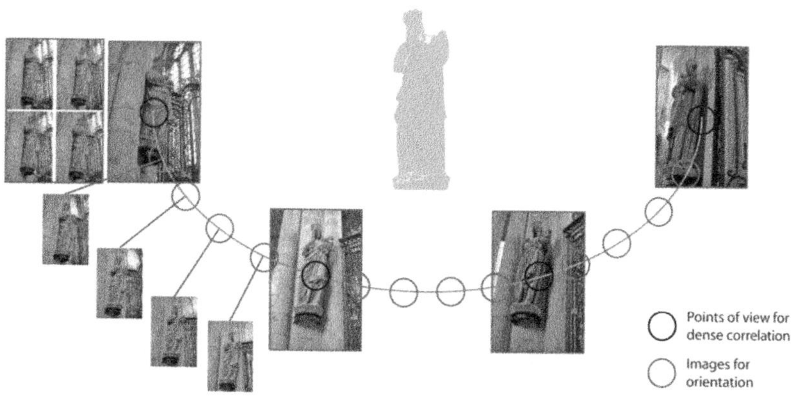

Figure 2.14. *Image survey strategy for a 3D object [HÉN 11]*

– *Combine potential viewpoints so that everything is expressed in the same coordinate system*

Imagining that four stereoscopic images have been taken to document a simple four-sided building, with a view to obtain four 3D point clouds. In order to guarantee that all the 3D point clouds can be used simultaneously, it is necessary to work within a single coordinate system. A simple way of achieving this is to take stereoscopic images and connect the viewpoints between them (Figures 2.14 and 2.15), while preventing two consecutive photo survey axes forming an angle above 10°. This is a constraint imposed by the scale invariant feature transform (SIFT) algorithm [LOW 04], which is very often used for the extraction of tie points (see section 2.4). In this way, there will be tie points between all the images of the block, thus enabling the calculation of the georeferencing of all images at once.

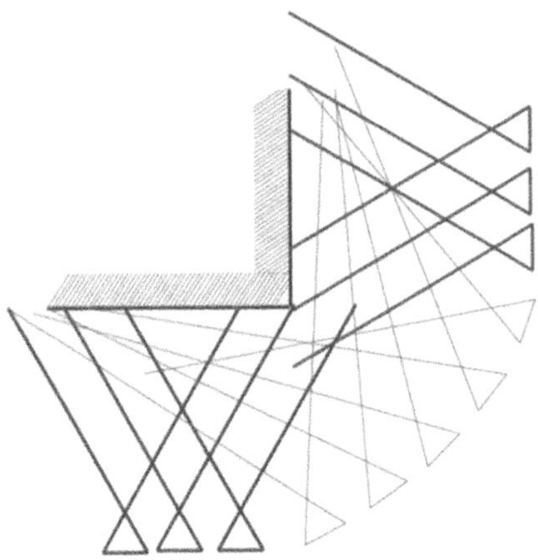

Figure 2.15. *Additional photos in order to express everything in a single coordinate system (as seen from above)*

– Take specific images in order to calculate the calibration of the camera

The fully automatic processes include the calibration of the camera in the orientation of images. However, certain configurations are partially or completely unsuitable for these processes: flat area, parallel photo survey axes and long focal length [FRA 13]. Thus, it is preferable to calibrate the camera on several images taken especially for this purpose, over an area with both texture and depth. For a given camera, any new combination "zoom + focus" theoretically requires a new calibration.

– If possible, overlay the lateral stereoscopic bases with perpendicular stereoscopic bases to make the matching algorithms more reliable

Indeed, bearing in mind that the search for the best homologous detail is carried out along a parallel line at the stereoscopic base (see section 2.5.2), it is useful to have overlapping images in a different stereoscopic configuration, in order to discriminate between the ambiguities among all the pixels contained in the objects parallel to the base. This recommendation is the reason for the cross-shaped image surveying (see Figure 2.16) where a central image is taken, followed by two images, one on each side, and two more, with one above and one below.

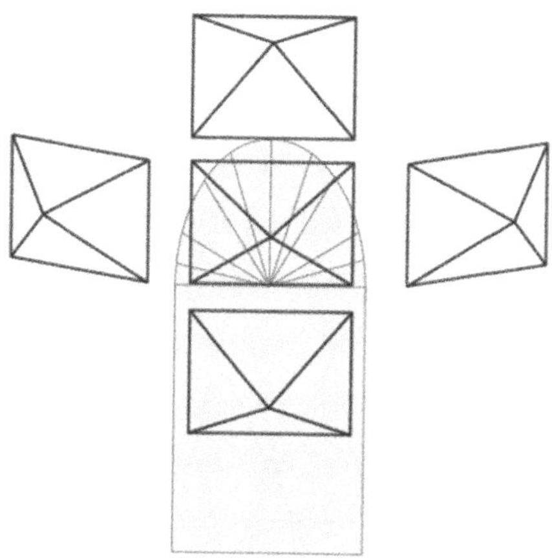

Figure 2.16. *Diagram of cross-shaped image surveying on the upper part of an entrance (five images)*

– Overlay images taken with a long focal length lens with wide angle images

The orientation of an image survey with long focal length lenses (> 100 mm) is a notoriously more delicate process than orientating images with a normal lens (the intersection of homologous perspective rays in these images is ambiguous, see Figure 2.17). To deal with this problem, surveyors advise overlaying more traditional images (taken with a normal- or wide-angle lens), which can be used to create a photogrammetric resection that will produce a number of control points to georeference the long focal length images with simple space resections (see section 2.4) [PIE 12]. Figure 2.18 show an example of such an image survey.

– Avoid empty space (sky and untextured areas), shiny surfaces and repetitive patterns

Given that untextured areas are not particularly conducive to automatic processing, on reasonably large areas it is possible to project texture that will help us to achieve the automatic orientation of the images, and sometimes even the 3D restitution.

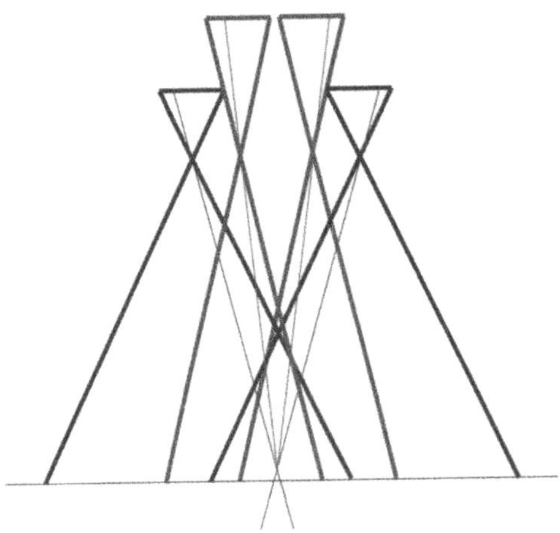

Figure 2.17. *The longer the focal length, the more the intersection of homologous perspective rays is ambiguous*

Figure 2.18. *Example of the joint use of images taken with different zooms (Arcade of Kings of Amiens cathedral)*

2.3.4. *Image surveying devices*

2.3.4.1. *Tripod, mast and pole*

To take static photos, it is advised to use a tripod, which will enable relatively long exposure times (especially when working with a mostly closed diaphragm, see section 2.3.1). In this case, a cable release can be used to avoid moving the camera during the opening of the shutter. If a cable release is not available, the camera can be used in delay mode. Obviously, we can say that the stability of the ground must be checked before starting the image survey on a tripod with very long exposure.

Certain configurations (inaccessible notable details) require the use of a mast or pole (Figure 2.19), to which the camera can be attached. It must be remotely controlled, often with the use of a computer, linked via a wireless network or a cable. Handling a pole of around 10 m in length is obviously more complicated than a traditional tripod for reasons of safety and its burdensome nature.

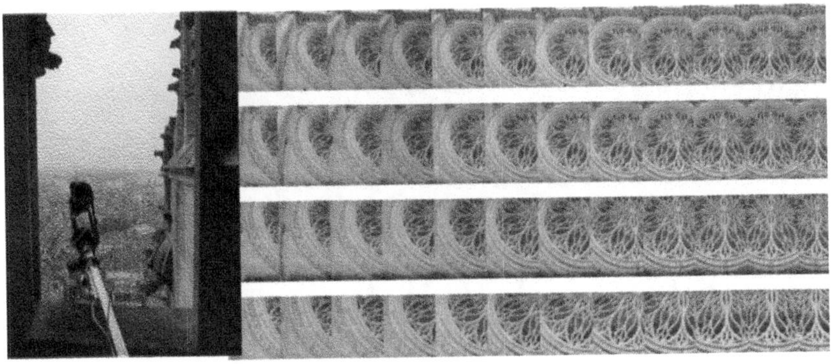

Figure 2.19. *Use of a pole and the highly redundant resultant image survey [HEN 12]*

2.3.4.2. *Panoramic head*

It can also be useful to use a panoramic head, which can produce wide coverage images at a very high resolution [DAN 13]. When working in this way (with tripod, pole with or without panoramic head) it is possible to combine images acquired at different exposure times in order to obtain the best exposure at each point (High Dynamic Range (HDR) fusion, by postprocessing [HAP 09]).

2.3.4.3. *Mobile acquisitions*

For reasons of efficiency (e.g. on large sites) or inaccessibility to certain areas (upper parts of structures), it is sometimes necessary to attach a camera to a car or aircraft. Some rules to be respected for all types of mobile acquisition can be provided.

Generally speaking, homogenous lighting is preferred, if possible with cloud cover, mainly to avoid too dark shadows, which can become even more problematic as they move while the image survey is being carried out. Regarding the camera, it must allow a rate which is sufficient for capturing very high resolution stereoscopic images (at least around 10 images per second) and must be controllable at a distance. On the majority of mass-market cameras, this rate is incompatible with the raw image recording format, which is larger than the compressed formats. Professional aerial cameras (those which are specifically designed for the acquisition of stereoscopic images for mapping) are often fitted with a forward motion compensation system to avoid any blur caused by movement of the camera during the opening of the shutter. On mass-market cameras where this device is not fitted, very short exposure times would have to be used.

2.3.4.3.1. *Acquisitions from aircrafts*

It is possible to carry out aerial image surveying for those with a modest budget by small airplanes, gliders or Unmanned Aerial Vehicles (UAVs), in order to access the upper parts of structures. Extra attention will be given to the weather conditions, especially wind, for such aircraft acquisitions.

UAVs are increasingly being used to survey outstanding sites because they enable very high resolution images at a relatively low price, with numerous viewing angles, on areas that are inaccessible from the ground [ZHA 11a]. These (Figure 2.20) can be helicopters, fixed-wings or airplanes (not to mention simple kites or balloons, which do not have their own means of propulsion and are therefore unable to follow a pre-established flight plan [EIS 09]). Helicopter-type UAVs are adapted for the photographic coverage of façades, or a general coverage for any area that requires stationary flight, while fixed wings or airplanes are used to fly over and photograph larger areas.

Regardless of the model, the use of a UAV requires prior steps to have been taken (insurance, flight authorization and pilot qualification), according to the situation which depends on the maximum distance of UAV and pilot, altitude of flight and type of area involved (populated or not). Whether the camera is fitted to a small piloted airplane, glider or UAV, a solid attachment system will be required where it can be operated at a distance (radio signal, infrared and cable), and have it validated by the International Civil Aviation Organization (ICAO)[4].

Figure 2.20. *Examples or aircraft and UAVs. a) Autogyro (the camera is fixed to a bar carried by the passenger), b) helicopter-type UAV and c) airplane-type UAV (sensefly[5])*

2.3.4.3.2. *Ground-based mobile acquisitions*

Even on the ground, productivity constraints mean that the digitization process may occasionally use mobile devices: vans, cars, carts, etc. Constraints on the high-speed rate of images are, in theory, less strict than for acquisitions by aircraft because it is easier to travel slowly without jeopardizing the stability of the camera holder. However, in practice, for reasons of cost effectiveness, cameras with high frame rates are also used. Inevitably, images from these devices will not be of the same quality (sharpness and lack of occlusions) as images acquired from a stationary position. Nevertheless, they can be sufficient for the survey of an entire district where the rendering is not the main objective. Although this type of acquisition has proved its worth, especially in whole towns [PAP 12], several

4 http://www.icao.int/Pages/FR/default_FR.aspx.

5 https://www.sensefly.com/home.html.

initiatives are underway to produce more focused surveys with small vehicles that are able to go beyond public areas and possibly enter buildings[6].

2.3.4.4. *Video mode*

Time can be saved on the acquisition of intermediate images (e.g. images linking different viewpoints, or wide-angle images taken to help the orientation of images taken with long focal length lenses) by carrying out image surveying in video mode. This video can then be converted into still images, which are then processed in the normal way. Admittedly, these images have lower resolution, and the process sometimes produces a number of blurred images, but they are not used in creating the final product [SAM 13].

2.3.4.5. *Stereoscopic cameras[7] and time-of-flight cameras*

Some equipment contains two or three cameras in order to produce stereoscopic images in a single click. The distance between the lenses is similar to the interpupillary distance (around 7cm). The exact synchronization of different lenses is not an essential factor in surveying outstanding sites (as it can be for the 3D modeling of moving surfaces). The interest in this type of material for surveying outstanding sites is to produce pairs or triplets of images at a known stereoscopic base, which helps to give a scale to the model [AHM 13]. However, since the stereoscopic base is small, it is not possible to create too much distance between it and the site being surveyed.

Time-of-flight cameras are able to provide a distance measurement for each pixel and can record up to 50 3D images per second. The size of the sensors is again rather small, but applications in the field of architectural surveying may soon be set to emerge [RIN 13].

---

6 http://www2.isprs.org/commissions/comm1/wgs.html#icwg_1-5a.

7 http://en.wikipedia.org/wiki/Stereo_camera.

2.3.5. *Survey control data*

At the time of image surveying, control points featured in the chosen referencing system (the global system) are earmarked, which will be used to georeference the photogrammetric products. These points can be natural details or artificial targets. Rules established for traditional production (block orientation – aerotriangulation – of standard images) mean that it is necessary to have a known point in three dimensions in each corner of the image block, as well as one point per stereoscopic pair. Absolute accuracy of the orientation is directly dependent on the control points used: measurements taken from targets will give better final accuracy than those taken from poorly defined natural points. It is estimated that the point accuracy is subpixelar on targets, pixelar on symmetrical marks and joint intersections and up to several pixels on rounded stone edges.

Although the equipment specifications for control points of traditional image surveys are well known, there are still few studies on the equipment necessary for redundant images in order to obtain the quality desired. However, since redundant image blocks are extremely rigid, a reduced amount of equipment should be more than sufficient (one point at each corner of the block, and one point every two or three pairs).

2.4. Image orientation

2.4.1. *Principles*

When the stereoscopic images are acquired, they are processed by a software chain in order to produce, more or less automatically, 3D point clouds, orthoimages or a 3D metric drawing. Whichever of these products is required and software is used, the first step is the orientation of the images. Evidently, fully automated tools that do not allow any interaction once the images are uploaded process this step like a black box, but implement exactly the same principles as the most expert tools.

The orientation of images aims to calculate the position (*S*) and orientation (*R*) of the cameras at the time when the image was captured, and in certain applications, the geometric characteristics of the camera used (autocalibration).

Direct georeferencing consists of recording the position and orientation of the camera at the time of the image being taken by external devices (e.g. GNSS antenna and inertial navigational system). In this case, photogrammetric orientation is not compulsory, but inertial navigational systems that enable the user to obtain the required accuracy for later photogrammetric applications are now exceedingly expensive. Only projects involving aircraft for large sites would benefit from direct georeferencing. Moreover, since inertial navigational systems need to be used with a GNSS receiver to compensate for any derivation, it is very difficult to use them indoors.

Photogrammetric orientation by image blocks, which still regard the overwhelming majority of outstanding structure survey sites, consists of a phase involving the measurement of tie points (visible details on several overlapping images) and possibly ground control points, or elements which help us to establish the scale of the final product, and then a calculation phase. The tie points are used to connect one image to another, either for a relative orientation of the images (here, this refers to global relative orientation to each other, without introducing scale or orientation yet), or to reduce the requirements for control points in the absolute orientation process (using the global coordinate system). Traditionally, a minimum of three tie points at the nadir of each image is chosen, as each point is then measured on all images where it is visible.

Figure 2.21. *Nadirs and tie points*

The orientation of a single image is called a space resection and is based on measurements of at least three ground control points.

The orientation of a block of images is calculated by resolving systems of equations based on fundamental formulas of photogrammetry (condition of collinearity and coplanarity in a stereoscopic pair). As a reminder (see section 2.2.4), the equation of collinearity: $m = \phi(M)$ describes the alignment of a point, M, with the camera position S, and the image of M on the image plane m (see Figure 2.6). The equation of coplanarity, based on the cancelling out of the triple product made of the stereoscopic base and two homologous perspective rays is $\overrightarrow{S_1S_2}.(\overrightarrow{S_1m_1} \wedge \overrightarrow{S_2m_2}) = 0$. This describes the fact that the two homologous perspective rays are coplanar (Figure 2.22). When this cancelling out is achieved for a sufficient number of tie points then the image pairs are correctly oriented relative to each other.

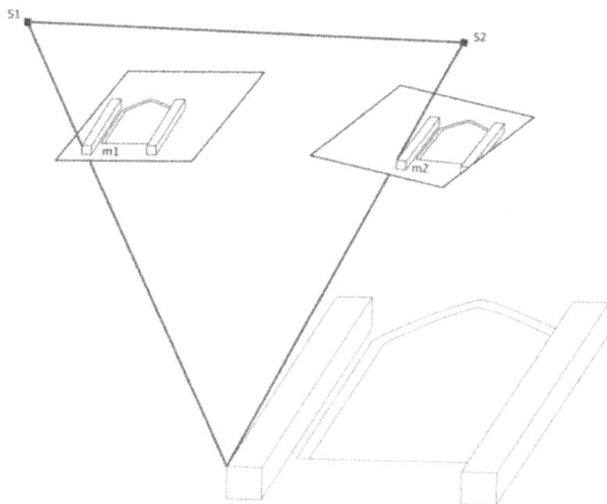

Figure 2.22. *Coplanarity of the base with two homologous perspective rays*

The steps of orienting an image block are progressive: the strips are first calculated using the relative orientations of the pairs which they consist of (thus implementing the coplanarity equation and 3D similarity transformations). They are then oriented against each other by once again applying 3D similarity transformations (seven parameter transformations, consisting of a rotation, a translation and a scaling) calculated on a minimum of three non-aligned points shared by the two strips. Finally, if the control data is available, the block at this stage is transferred to the global system by applying a 3D similarity transformation calculated on at least three control points. A thorough adjustment of all measurements (for tie and control points) is ultimately calculated to adjust the orientation of the images as best as possible. To do this, collinearity equations are used, linearized around an initial solution put forward, calculated by the previous steps (sequence of pairs then strips). During this step, known as "bundle adjustment", measurements are weighted to take account of accuracy and various orders of magnitude of input data.

2.4.2. *Automation*

The full, high resolution survey of outstanding sites by photogrammetry involves a large number of images which are varied in terms of the photographic lenses used in order to cope with all possible configurations (short focal length lenses in case of lack of distance from structure, long focal length lenses for areas of interest which are farther away, etc.). Therefore, it is essential that the image orientation phase is as automated as possible.

2.4.2.1. *Automation of tie point measurement*

Measuring tie points is a long process when done manually. Fortunately, in most cases nowadays it is done automatically. Currently, the most reliable methods are based on the following principle:

– detection of points of interest, whose position is not ambiguous, which can be identified in a clear and repetitive way on all the images in the block;

– matching: the points of interest which represent the same detail on several images are given the same identifying reference.

The point of interest detectors frequently used are scale-invariant feature transform (SIFT), affine SIFT (ASIFT), speeded up robust features (SURF) and maximally stable external regions (MSER) [BAR 10]. The SIFT algorithm for instance, since it is able to detect common points of interest between overlapping images even if there are differences of scale or rotation, is particularly suitable in the context of close-range photogrammetry. However, it is less reliable in terms of viewpoint changes, which explains the often-stated constraint to avoid separating two consecutive photo survey axes by more than 10° in a stereoscopic image survey. It is to be noted that ASIFT, a variant of SIFT which is more demanding in terms of calculations, is not sensitive to affine transformations between images. Some software chains use SURF, optimized to speed up calculation time, or MSER, which is a detector of regions, and is not sensitive to viewpoint changes.

The matching phase can be carried out by measuring similarities between photometric patches extracted from around the points of interest (see also section 2.5.2) or by using more evolved descriptors. The matching of SIFT points, for instance, can be done by using descriptors of radiometry gradients around the points of interest (matching is made between the points which are closest in space at 128 dimensions of these descriptors). Matchings of pairs are then spread to obtain multiple matchings, since tie points in photogrammetry become more useful (and help the reliability of the reconstituted image block) when they are measured over a large number of images. Taking proximity into account can help us to avoid errors, e.g. two neighboring points cannot be matched to two points that are a long distance away. Occasionally, we may decide to refine the accuracy of the multi-image matching by automatic correlation.

The amount of tie points obtained in this way, of course, depends of the content of the images: the more textured and close together the images are, the more tie points will be captured. Algorithms such as SIFT easily provide thousands of points per image, although traditional photogrammetric tools only require a handful of them. Some software tools, such as Microsoft Photosynth[8], also provide tie points reconstituted in 3D following the orientation of images, as 3D point clouds. Possessing comfortably more points than the required minimum is not an impracticable luxury: densely packed points across the whole surface of the images help to bolster the strength of the autocalibration.

8 http://photosynth.net/.

Nevertheless, it is important to be able to effectively filter out inaccurate points. This is how the RAndom SAmple Consensus (RANSAC) estimation method [FIS 81] proved its worth [BAR 10].

Finally, on some modern buildings which lack well-defined regular details (e.g. concrete bridge piles), it is possible to use more basic elements than points, such as segments, in order to link images to each other [HOT 75, ZHA 11b].

2.4.2.2. *Semi-automation in measuring ground control points*

The use of coded targets in the field enables automated capturing of control points. Some software tools also enable us to facilitate the measurement of more traditional control points (standard targets and natural points of details). They recommend starting by measuring three well spread-out control points, after a global relative orientation calculation, in order to establish absolute orientation, giving accurate values for cameras' positions and orientations, which in turn supply the image formulas. Every remaining control point can, thus, be projected in the images where it is visible, very close to its correct position. The operator needs then to refine the point to its exact location [FON 12].

2.4.2.3. *Exterior orientation of images and autocalibration*

In favorable cases, the current software chains are able to take all input images and choose the best strategy for putting them together. These strategies consist of, first, processing wide-angle images (recognized by EXIF metadata). Images taken with a longer focal length are notoriously more delicate to orientate. If the usual strategy is unsuitable (see section 2.4.1), common tie points between shorter focal length images which have already been oriented and images to be processed will provide control points to carry out space resections (resolution of unknown (S) and (R) solely from the collinearity equation) [PIE 12].

It is often possible to integrate the calibration of the camera into the orientation calculation. In this case, the calculations described beforehand are done using internal camera parameters which are already known, or at least with approximate parameters, estimated using EXIF data. It is assumed that there is no distorsion, that the principal points are at the geometric center of the sensor and that the principal distance is as provided by the manufacturer. When the calculation of the orientations of the images converge, the parameters of the camera are re-estimated to improve the overall result (especially the intersection of homologous perspective rays). However, it is important to reiterate that at the time of the acquisition it is wise to earmark images specifically for the calibration, on an appropriate area (convergent images of a textured area with depth).

2.4.3. *Strategies for the orientation of all the images of a site*

It is common to divide the processing of the images into different parts of the area of interest. As a result, images of the façades, roofs, statues and so on are processed separately. For each section, the first step is a global relative orientation calculation, hence without control points. The result can be qualified by the average distance between the homologous perspective rays, often expressed in terms of pixels. In case of difficulties, where the calculation diverges, or converges toward a value which is too high, the input data should be verified (camera parameters, any images that may have been rotated or cropped), and then a smaller block is recalculated, and images are added step by step. If problems persist, the flaw may lie in a poorly calibrated camera, misconfigured image survey, a difficult study object (lack of texture) or a combination of several of these. When the image block has undergone global relative orientation, several (at least three) control points are measured and an absolute orientation is calculated to check that they are suitable (correct numbers and appropriate coordinate system). Finally, all the points are measured and the calculation is completed (Figure 2.23).

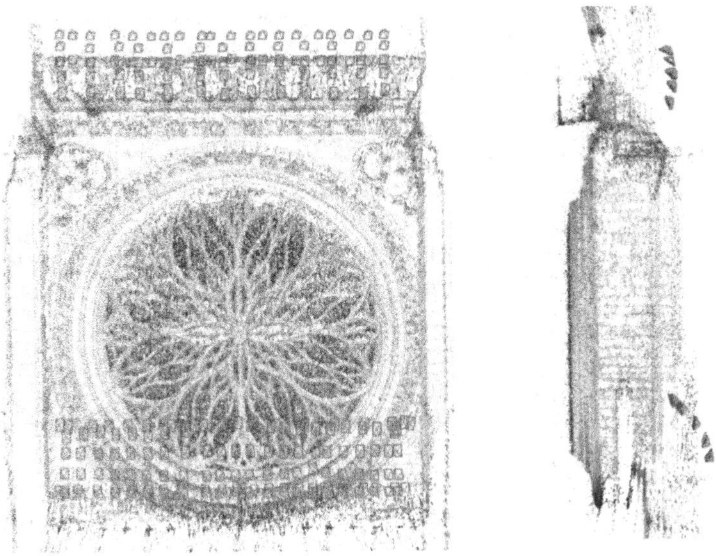

Figure 2.23. *3D Point cloud showing the position of cameras and tie points (rose window of the Western façade of Amiens cathedral)*

When the image survey is suitable for it, (when images that have been taken are able to connect different sections to each other) attempts can be made to calculate the orientation of several sections in a single calculation, in order to optimize the relative positioning of all the images, without which we would be solely dependent on control points. For large structures, it is still difficult to orient all images at the same time for logistical reasons connected to the substantial volume of data to process.

2.4.4. *Qualifying the orientation of images*

The orientation of images needs then to be validated, since the quality at this stage affects the entire subsequent production process. When possible, several aspects should be verified:

– the average distance between homologous perspective rays (around 1/3 of a pixel, to ensure subsequent high-quality processing, whether done manually or automatically);

– residual to any existing ground control points and other check points (recorded on the reference system but not used for the image orientation calculation) in accordance with the prior typical deviation of control points;

– an acceptable proportion and repartition of preserved points.

2.5. Production of 3D point clouds from images

2.5.1. *Definition and use of 3D point clouds*

A 3D point cloud is a collection of points described by coordinates *(X,Y,Z)* in a coordinate system, and usually by a color (Figure 2.24). There is not necessarily any topological relation between the various points. The 3D point clouds can be useful in the context of outstanding site surveys because they represent, on the one hand, a virtual and metric aid for operators during the restitution and, on the other hand, a dense description of the objective form of the areas photographed (as opposed to 3D models extrapolated from a handful of points and a construction hypothesis) [HÉN 10]. The processes for point clouds to undergo in order to fully benefit from them are described in Chapter 5.

Figure 2.24. *3D color point cloud (Lincel chapel)*

2.5.2. *Principle for the production of 3D point clouds by dense image matching*

A very simplified version of the principle for producing 3D point clouds by dense image matching is the following: an image is matched with its neighbors, generally on a pixelar level (in other words, for each pixel of an image its homologous must be found in the neighboring images), but possibly even on a subpixelar level. By using the intersection of perspective rays from matched points, this enables the creation of a 3D point cloud with a resolution close to the average resolution of the images used (see Figure 2.25).

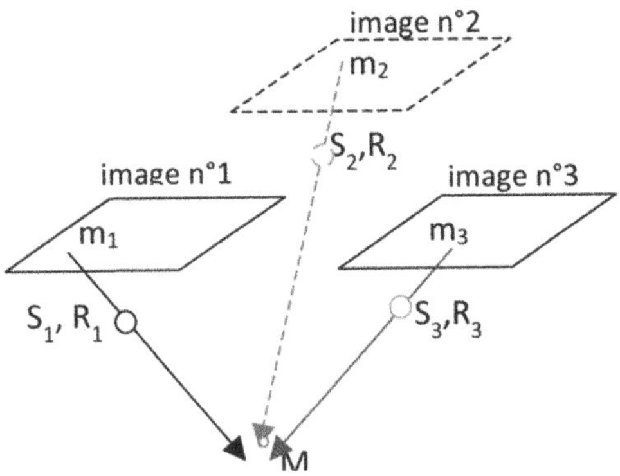

Figure 2.25. *Construction of 3D points by intersection of homologous perspective rays*

2.5.2.1. *Similarity function*

The matching of images is generally carried out on patches (subsections of images) and consists of measuring their similarity, often given by the correlation coefficient, *cor*.

$$cor = \frac{\sum_{i=1}^{n}(x_i - \bar{x})(y_i - \bar{y})}{\sqrt{\sum_{i=1}^{n}(x_i - \bar{x})^2 (y_i - \bar{y})^2}}$$

– $\vec{x}$ is a vector at n dimensions; its coordinates are simply the radiometric values contained in the reference patch;

– $\vec{y}$ is a vector at n dimensions; its coordinates are the radiometric values in the patch being compared to the reference patch;

– $\bar{x}$ is the average of coordinates of $\vec{x}$, while $\bar{y}$ is the average of coordinates of $\vec{y}$;

– n is the total number of pixels in the reference patch (for example $n = 9$ for a patch of 3×3).

It is easy to demonstrate that *cor* is between –1 and 1. When *cor* equals 1, the patches are proportional (or identical, apart from light levels).

2.5.2.2. *Image matching from the "image" space or from the "object" space*

Correlators generally allow work in the "image" space. This means that they provide, for all the pixels of a given image, known as the "master image", homologous pixels in the neighboring images. The pixels themselves are not matched, but the patches centered on the pixels are. The search is carried out for all image pairs possible, in the epipolar geometry of corresponding couples (Figure 2.26): for a given pixel of the reference image, epipolar geometry limits the search area of the homologous point to a straight line (assuming there is no optical distortion) on the neighboring image. This straight line is the intersection of the plane of this image with the plane which passes through the base and the perspective ray from the point to be matched on the reference image.

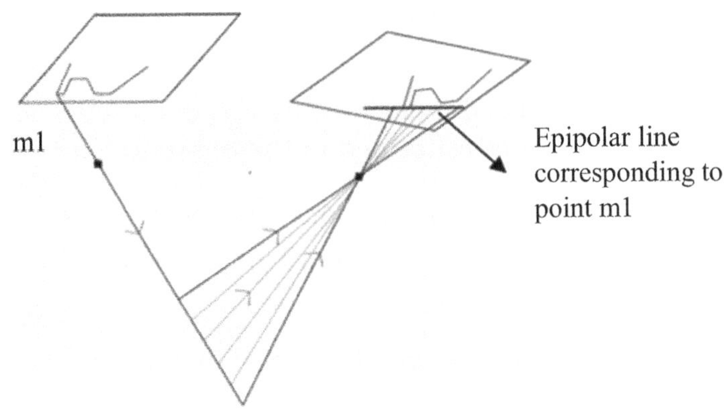

Figure 2.26. *Use of epipolar geometry for matching in the "image" space*

The result of the matching is a disparity map or a parallax map. The disparity or parallax in a pixel of the image is the deviation in the position of the homologous pixel on the neighboring image. The disparity map can be directly converted into a depth map (Figure 2.27): this is an image of the same dimensions as the reference image, in whose pixel values the depth of the scene on the perspective ray can be seen. This output geometry is adapted to truly 3D structures, such as columns, stairs and statues. To completely restitute these structures, a depth map should be produced from each viewpoint (front, back, above, below, etc.).

Figure 2.27. *Master image and depth map (Arcade of kings, Amiens cathedral)*

Some correlation algorithms also facilitate the exploration of the object space. In this case (see Figure 2.28), the object (or part of the object) is sliced into pixels that are projected at varying depths into images that show it. For each pixel in the object space, the depth which provides the most similar patches will be kept (based on the hypothesis that an object is seen in the same way on different images it appears in, which is only guaranteed in the case of truly isotropic surface). If the objects being processed are mainly planes (façades, ground seen from above, etc.), the consequent depth map from the dense matching phase can be an image placed on a chosen plane of the object (often the object's average plane is chosen), in the pixels of which the distance to the plane can be found. This output geometry can be adapted to cope with objects that have different forms, such as domes, which will initially be modeled in hemispheres.

Often, to reduce calculation time and guide the result, the working area is limited by masks or silhouettes, from manual drawing or automatic extraction.

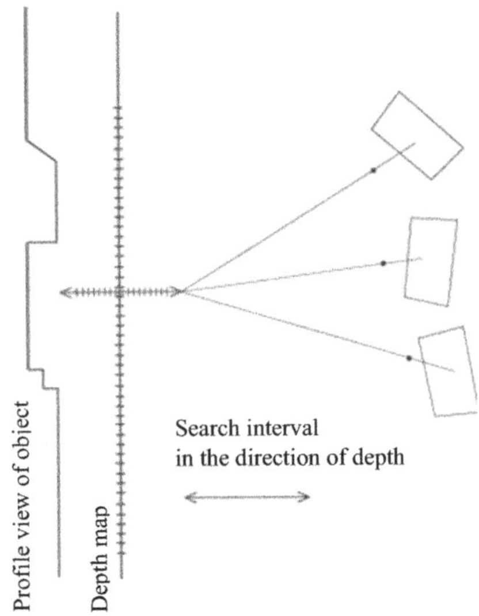

Figure 2.28. *Image matching from the object space*

2.5.2.3. *Output data*

The resultant data produced consist of depth maps, which can be converted into 3D point clouds. The various clouds that cover the area are logically placed adjacently to each other if the image orientation is carried out in a global way with a single coordinate system, but the joint areas must undergo specific processing. These 3D point clouds are naturally colored by the images used for the matching (indeed, any 3D point in the cloud emerging from the intersection of several perspective rays can take on the color of the pixel that one of these rays came from).

2.5.2.4. *Typical pitfalls of image matching*

Whatever the approach (matching in image or object space), the matching of homologous pixels can fail on homogenous areas or repetitive details in images. Moreover, when working on the image space, linear objects parallel to the stereoscopic base (parallel to the direction of the research) can cause difficulties which supplementary images can help to resolve, as seen in section 2.3.3. Finally, the size of the patches is decisive: small patches are more suitable for the restitution of discontinuities, but cause a more noised overall result. Many image-based 3D reconstruction methods are available [SCH 02, SEI 06], which show similar levels of ingenuity in order to overcome the pitfalls mentioned above and optimize the final result.

2.5.3. *Optimization*

2.5.3.1. *Hierarchical approach*

To optimize the result in terms of speed and quality, the search for the best homologous points is done in multiresolution (using pyramids of images): first, work is done with a highly subsampled set of images, allowing the production of a rough layout that can be refined step-by-step on images with gradually improving resolution (Figure 2.29) and also enables a reduction of the research area for each pixel.

2.5.3.2. *Global method and optimization*

In reality, to rebuild a true-to-life surface, we cannot simply be satisfied with placing obtained results adjacently in an independent way pixel after pixel. A global method needs to be used, which has the advantage of processing the entire surface all at once, and the disadvantage of being extremely cumbersome in calculation time. Nevertheless, optimizations are possible, such as dynamic programming or graph cutting methods. These methods consist of looking for the minimum of an energy function made up of one part controlling the similarity between images and another part for the surface regularization to be reconstructed [PIE 06, HIR 08].

Figure 2.29. *Pyramid of images and depth map becoming more and more accurate*

2.5.3.3. *Influence of parameters*

Evidently, the final result depends not only on the input images (geometry of the image survey and image quality), but also on the parameters used for the calculations for the depth map. The results vary according to the size of the correlation patches, the number of images in the pyramid and the factor of surface regularization. Depending on the software used, these parameters may be changeable or fixed by default.

2.5.3.4. *Scalability*[9]

It should by emphasized that dense image matching algorithms have long been the subject of literature [BAI 97], but have only become effective in the past few years. This is due to the image quality from digital cameras (compared to scanned films which were used in the initial tests) and the power of points of interest detectors, such as SIFT, which enable complete automation of the photogrammetric process as soon as the image surveying is completed and hence is able to process a significant number of images.

The "Middlebury[10]" challenge showed that a large number of algorithms for 3D reconstruction, which came mainly from the "computer vision" community (where the term *Structure from Motion* (SFM) is commonly used), are now operational on small objects. However, their ability to process real-size sites is variable, and they often require algorithmic adjustments [VU 09] and organizational and logistical amendments.

Scalability is used when proven methods on small test sets need to be implemented on real-sized data. The processing constraints of digital data described in articles on

---

9 http://en.wikipedia.org/wiki/Scalability.

10 http://vision.middlebury.edu/mview.

big data[11] are in vogue for surveying projects of structures that are completely done by photogrammetry. Issues regarding the complexity of sites, perspective, desired resolution and available photographic material mean that the number of photographs that need to be taken for complete digitization cannot be predicted. Moreover, it can be expected that intermediate data will be added to the original images (tie points, pyramid of images, depth maps, etc.), as well as safety backups. Therefore, data processing requires proper organization:

– powerful computer units, highly equipped with storage space, connected to a network to allow calculation sharing (*cloud computing* [LEY 09]);

– optimization of algorithms;

– parallelization of coding;

– graphic processing unit (GPU) programming (use of graphic processor to increase the calculation power of the computer) [ROT 13];

– in-database processing [CAN 13];

– standardization of data names according to the area of the structure covered;

– standardization of file names according to the stage of the process;

– mandatory production of quality metadata for each step in the process.

2.5.4. *Qualification of data produced*

Significant reconstruction errors are often linked either to inappropriate materials for image matching (materials lacking texture or overly shiny, pierced surfaces), or to poor

11 http://en.wikipedia.org/wiki/Big_data.

geometric configuration of the image survey (inappropriate camera, insufficient number of images and blurred images). These are easy to identify, directly on the 3D point cloud, or better still, on an image showing the relief of the scene brightened by an artificial light. However, the final result may contain more subtle errors (general noise, bad localization, local artifacts, etc.). Scrupulous inspection of the shaded relief images enables the detection of some of these, but is not able to quantify them. Only the quality of the absolute positioning can be evaluated with the help of ground control points.

Problem observed	Possible causes	Solution
Large residuals to ground control points	Incorrect control point, not detected during orientation Calibration/geometric stability problem	Redo the orientation of images
Reconstruction with holes	Inappropriate material → incorrect correlation coefficient Hidden areas	Project texture if possible. If not, there is no solution. If possible, add more images
Noised reconstruction	Poor image quality, incorrect configuration	Redevelop original images, do not use compressed images Change the parameters (regularization coefficient)
“Soft” reconstruction	Incorrect parameters	Change the parameters (image similarity coefficient)

Table 2.1. *Inventory of problems observed regarding automatic photogrammetric processing of images and possible causes*

2.5.5. *Advantages and limits*

Photogrammetry enables the almost automatic creation of dense 3D point clouds in color. For this reason, it is completely suitable for the surveying of outstanding sites. Its advantages are summarized below:

– inexpensive cost of material;

– heavily automated process, accessible to all;

– flexibility of photographic material used (from fisheye to telephoto lenses possibly attached to masts, poles or UAVs);

– adaptable accuracy (distance from the object of interest, focal length of lens and stereoscopic configuration).

However, its limits are the following:

– large volume of data to process, long calculation time;

– inability to process certain materials, such as those shiny or hollow in nature;

– difficulty in systematically qualifying the results obtained.

2.5.6. *Other approaches to automatic image-based 3D reconstruction*

The approach discussed in section 2.5.2 looks to produce one depth map per viewpoint. Its variations are presented in [SCH 02], especially the key issues: similarity measurements, surface regularization and optimization. Through this approach, the complete 3D reconstruction of an object or a site requires the merging of the depth maps, while processing the hidden areas. However, there are different "image-based" methods, discussed at length in [SEI 06], where the 3D scene is not an assembly of depth maps, but is modeled:

– by a collection of voxels: method of carving, which consists of removing from an initial volume of voxels which do not respect the similarity criteria in the images [KUT 00];

– by *level sets*: minimization of a collection of partial differential equations defined on a volume [FAU 98];

– or by evolutionary triangulation: densification by an approach of tetrahedrization based on the points of interest taking into account visibility constraints [VU 09, FUR 10].

The interest of these methods, compared to methods that create depth maps, is to provide a continuous surface of separation between the object and the empty space, but they are more suitable for small objects than for large sites (except the method of evolutionary triangulation, which is just as effective on small objects with significant discontinuities as on large areas). Potential users should not be discouraged by the multitude of algorithms for image-based 3D reconstruction. On the contrary, they are able to offer multifunctional software chains which are adapted to all configurations (including the processing of refractory areas by image matching) as demonstrated for instance in [REM 08].

2.6. 3D drawing by stereo or multi-image plotting

The traditional application of photogrammetry consists of manually drawing on a pair of images viewed by stereoscopy. This reliable technique has proved its worth on both aerial and ground-based images. It is described in this section along with measurements from a larger number of images. The methods used to directly process point clouds are discussed in Chapter 5.

2.6.1. *Definition and use of 3D drawing*

A 3D drawing is a model made up of isolated points, lines or polygons described by coordinates (X,Y,Z). It is able to represent the characteristic elements of the structure's geometry according to the project specifications. The concept of an appropriate wire-framed model is used, since it shows only that which has been requested. Indeed, unlike the point cloud, which although dense, will not give an exact rendering of a structure's edges, a manual drawing will provide the rendering as described in the specifications (for instance, the detail of an exposed wall's brick work or the arch of a façade).

2.6.2. *Principle of the production of a 3D drawing*

A 3D drawing can be created in multi-image mode, as with the Photomodeler[12] software. In this case, the operator independently positions the cursor on the desired area of detail in all overlapping images available. The process is guided by the geometric constraints arising from the stereopair (constraint of epipolarity, see Figure 2.26), and assisted by automatic correlation (see section 2.5.2).

The principle of photomodelization described in [DEL 09] is based on measurements of points in stereoscopic images, and possible extrapolation with the supposed parallelism and orthogonality of the architectural elements to be reconstructed.

It is also possible to create a drawing by viewing stereopairs in 3D. To do so, it is necessary to have stereoplotting equipment consisting of:

– a computer equipped with a graphics card which enables 3D visualization and the fluid manipulation of images;

– a 3D screen which allows two images to be displayed in stereoscopy;

---

12 http://www.photomodeler.com/index.html.

– stereoscopic glasses which allow the left eye to see only the left-hand image, and the right eye to see the right-hand image;

– a 3D mouse to move the cursor within the photographed scene;

– a trained operator.

The edge matching between neighboring pairs must be done with great care in order to provide a continuous final rendering.

The principle of stereoplotting with digital equipment is described in Figure 2.30:

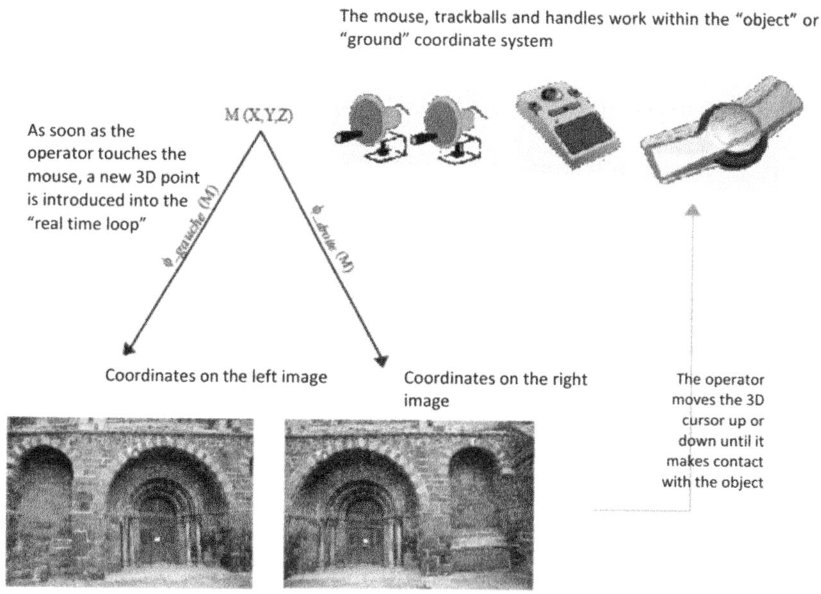

Figure 2.30. *Real-time loop in stereoplotting*

Effective tools enabling the combined use of automated orientation algorithms and traditional stereoplotting are yet to be created, while all the ingredients of such a tool are now available. These would allow the implementation of an image block without any intervention from the operator (as all chains of dense image matching do nowadays), and would also offer an interface to manually capture the required elements in specific stereopairs.

Intensive stereoplotting as was practiced before the digital era is now a rarity. However, it is still the most effective solution for accurately drawing specific elements on a structure (see Figure 2.31). It is worth reiterating this, because in the age of manifold images and automatic relief that we have been living in for several years now, it is easy to forget that occasionally it can be more direct to draw characteristic elements required onto the image rather than onto a 3D point cloud [ALM 10, HÉN 11].

2.6.3. *Qualification of 3D drawing*

It is assumed that the orientation of the images is of a satisfactory quality. For a long time, the potential of 3D restitution on stereopairs has been recognized. The fact that work is now done with digital images changes nothing: it is estimated that with images of normal quality, a trained operator can draw to an accuracy of half a pixel within the reference plane. The accuracy of the drawing in terms of depth should be balanced by the H/B ratio (distance from the scene to stereoscopic base). The specifications of the process must be clearly set out for the operator to avoid overgeneralizations or overdigitization. Contrary to automatic methods, this process can be applied to the majority of materials, even when they have little texture. However, the configuration of the pair supplied to the operator must be practicable (parallel axes in the image survey). If this is not the case, it is necessary to work on new images obtained by epipolar resampling (resampling which produces images whose lines correspond to the epipolar lines of the pair). The qualification of the drawing involves the completeness of the rendering, its suitability and accuracy. The process is difficult to automate and can give rise to cross-checks between different operators. Graphical output also enables the operator to evaluate what remains to be done, the quality of the lines (are they too simplified?) and the correct codification of the elements captured.

Figure 2.31. *Organ and cloister of choir of Seville Cathedral, drawn by stereophotogrammetry [ALM 10]*

2.7. The software offer in close-range photogrammetry

Traditional manufacturers of photogrammetric material now offer software tools, which are regularly updated with the latest advancements in research. However, they are mainly designed for the processing of large blocks of aerial images. Today, new tools produced through combining expertise of photogrammetrists and computer-vision engineers enable the processing of any kind of image survey configuration, including completely unstructured acquisitions made up of images of a single structure harvested from the internet. Their format is extremely variable, from free or even open-source software, to more traditional pay services, from locally installable applications with high numbers of variable parameters designed for use by experts, to web application where the user can simply upload a set of images and await the result.

2.8. Research prospects

Recent advances in photogrammetry, including the possibility of producing 3D color point clouds only from stereoscopic images, in the same way as laser scanners, come directly from the image processing research laboratory. Other areas, more or less connected to these tools are currently the subject of studies or research.

2.8.1. *Production in near-real time*

On the face of it, the production of data in near-real time is not a crucial requirement in the field of surveying outstanding sites, like it is, for example, for robotic applications. Nevertheless, being able to produce results in near-real time is desirable, on the one hand, to prevent data acquired from accumulating without being processed and, on the other hand, to react quickly in case of a flaw in the acquisitions.

2.8.2. *Data crowdsourcing*

Crowdsourcing is booming in all areas thanks to the Internet and especially social networks. The most well-known example in the field of geomatics is OpenStreetMap[13], a database, mostly of roads, which is updated by global positioning system (GPS) traces of its users [HEI 10]. For the surveying of outstanding sites, [SNA 06] have been demonstrating, for several years, that it is possible to model a monument using images harvested from the Internet. The opportunity is open to everybody to contribute to surveys of such sites, whether it be on a geometric or on a semantic level, by uploading images and annotations via a tablet or smartphone to dedicated servers. Theoretically, this resolves economic and even cultural problems, for instance in countries that are not able to invest in the inventory of their architectural heritage [HÉN 09, SYL 13]. This also gives rise to new computer-related questions (Which architecture to choose for fast transfer? Which data format to use for the greatest universality and for the quickest data processing?), and especially on a qualitative level (How can we control input data?).

2.8.3. *Automatic reconstruction of 3D vector models using images*

Finally, the prospect of automatically producing a 3D vector model has caused a stir among many researchers. The promising results obtained on aerial image surveys with a view to automating the production of urban databases is regularly published [CHE 04, HAA 10, HAM 11], sometimes using existing databases to improve the result [VAL 11]. Automatic extraction of views of façades, photographed in the street by a mobile mapping device, with or without a supplementary laser acquisition [PIN 11], as well as the

13 http://openstreetmap.fr/.

extraction of buildings using video [TIA 10] also appear to be possible. However, there are fewer results on vector mode reconstruction using images taken for a very high resolution survey of a given structure. There are indeed various obstacles, which make the process difficult to automate: generally less organized image surveys than with aerial or mobile digitization, potentially less regular characteristic lines to reconstruct caused by short distances from the structures and any possible deformations, as well as hidden sections [BAR 10]. As shown in Chapter 5, reconstruction is more focused on 3D point clouds and will jointly use the radiometric information provided by the images and the geometric information of the 3D point clouds.

2.9. Bibliography

[AHM 13] AHMADABADIAN A., ROBSON S., BOEHM J., *et al.*, "A comparison of dense matching algorithms for scaled surface reconstruction using stereo camera rigs", *ISPRS Journal of Photogrammetry and Remote Sensing*, vol. 78, pp. 157–167, April 2013.

[ALB 11] ALBERTZ J., "Albrecht Meydenbauer – pioneer of photogrammetric documentation of the cultural heritage", *Proceedings of the 18th CIPA Symposium,* Potsdam, Germany, 2011.

[ALM 10] ALMAGRO A., "Photogrammétrie pour la recherche architecturale: vingt ans d'expérience à l'École d'Études Arabes", *Revue Française de Photogrammétrie et de Télédétection*, vol. 196, pp. 10–20, 2010.

[BAI 97] BAILLARD C., Analyse d'images aériennes stéréoscopiques pour la restitution 3D des milieux urbains. Détection et caractéristique du sursol, PhD Thesis in Signal and Image Processing, Ecole Nationale Supérieure des Télécommunications de Paris, France, 1997.

[BAR 10] BARAZETTI L., SCAIONI M., REMONDINO F., "Orientation and 3D modeling from markless terrestrial images: combining accuracy with automation", *The Photogrammetric Record,* vol. 25, no. 132, pp. 356–381, December 2010.

[BON 58] BONNEVAL H., "Utilisation de la photogrammétrie aux levés architecturaux. Application au temple d'Abou-Simbel (Haute-Egypte)", *Communication de l'Institut Géographique National au Congrès des Sociétés Savantes*, Paris, 1958.

[CAN 13] CANNELLE B., Estimation de pose centimétrique sur un corpus de plusieurs milliers d'images terrestres acquises depuis la rue, PhD Thesis in Geoinformation Sciences, Université Paris Est, 2013.

[CAR 78] CARBONNELL M., Le fil des Pierres: Photogrammétrie et Conservation des Monuments, Département des Relations Publiques de la Société Kodak-Pathé, October 1978.

[CHE 04] CHEHATA N., PIERROT-DESEILLIGNY M., STAMON G., "Génération de MNE hybrides (raster/vecteur): optimisation à base de flots de graphes contrainte par des primitives", *Revue Française de Photogrammétrie et de Télédétection*, vol. 176, pp. 30–41, 2004.

[DAN 13] D'ANNIBALE E., TASSETTI A.N., MALINVERNI E.S., "From panoramic photos to a low-cost photogrammetric workflow for cultural heritage 3D documentation", *Proceedings of the 24th CIPA Symposium*, Strasbourg, France, 2013.

[DEL 09] DE LUCA L., *La Photomodélisation Architecturale. Relevé, Modélisation et Représentation D'édifices à Partir de Photographies*, Eyrolles, 2009.

[DEN 30] DENEUX H., *La Métrophotographie Appliquée à l'architecture,* Paul Catin Editeur, p. 82, Paris, 1930.

[EIS 09] EISENBEIß H., *UAV Photogrammetry,* Institut für Geodäsie und Photogrammetrie and der Eidgenössischen Technischen Hochschule, IGP Mitteilungen, vol. 105, Zürich, 2009.

[FAU 98] FAUGERAS O., KERIVEN R., "Variational principles, surface evolution, PDE's, level set methods and the stereo problem", *IEEE Transactions on Image Processing,* vol. 7, no. 3, pp. 336–344, 1998.

[FER 04] DE FERRIÉRES M., *Éléments de Technologie Pour Comprendre la Photographie,* Editions VM, 2004.

[FIS 81] FISCHLER M.A., BOLLES R.C., "Random Sample Consensus: A paradigm for model fitting with application to image analysis and Automated Cartography". *Communication Association and Computing Machine*, vol. 24, no. 6, pp. 381–395, 1981

[FON 12] FONTAINE C., "Création d'une Interface pour le Positionnement Automatique des Points d'Appui", Rapport de stage du cycle ingénieur 2ème année, Ecole Nationale des Sciences Géographiques, 2012.

[FRA 13] FRASER C., "Automatic camera calibration in close-range photogrammetry", *Photogrammetry Engineering and Remote Sensing*, vol. 79, no. 4, pp. 381–388, April 2013.

[FUR 10] FURUKAWA Y., PONCE J., "Accurate, dense, and robust multiview stereopsis", *IEEE Transactions on Pattern Analysis and Machine Intelligence,* vol. 32, no. 8, pp. 1362–1376, August 2010.

[GRU 94] GRUSSENMEYER P., "Photogrammetry for the study and safeguard of historic sites and monuments", *Workshop on New Methods and Techniques for the Preservation of Historic Sites and Monuments, Cairo,* Ecole des mines, Laboratoire de mécanique des terrains, Nancy, France, 4–8 December 1994.

[HAA 10] HAALA N., KADA M., "An update on automatic 3D building reconstruction", *ISPRS Journal of Photogrammetry and Remote Sensing*, vol. 65, pp. 570–580, 2010.

[HAM 11] HAMMOUDI K., DORNAIKA F., "A featureless approach to 3D polyhedral building modeling from aerial images", *Sensors Journal, 10 Years Sensors – A Decade of Publishing*, vol. 1, pp. 228–259, 2011.

[HAP 09] HAPPA J., MUDGE M., DEBATTISTA K., *et al.*, "Illuminating the past – state of the art", *Proceedings of the 10th International Symposium on Virtual Reality*, Archaeology *and Cultural Heritage*, VAST – State of the Art Reports, Valetta, Malta, 2009.

[HEI 10] HEIPKE C., "Crowdsourcing geospatial data", *ISPRS Journal of Photogrammetry and Remote Sensing*, vol. 65, no. 6, pp. 550–557, 2010.

[HÉN 07] HÉNO R., EGELS Y., "Architectural patrimony management in Yemen", *Proceedings of the 21st CIPA Symposium*, Athens, Greece, 2007.

[HÉN 09] HÉNO R., "Towards a sustainable project for architectural heritage conservation in Yemen", *Proceedings of the 22nd CIPA Symposium*, Kyoto, Japan, 2009.

[HÉN 10] HÉNO R., *et al.*, "Archéologie et photogrammétrie: les nouveaux potentiels", *Géomètre*, vol. 2075, pp. 26–40, November 2010.

[HÉN 11] HÉNO R., CHANDELIER L., "Complementarity of terrestrial laser scanning and DSM from aerial photography: the example of the cathedral of Amiens", *Proceedings of the 23rd CIPA Symposium*, Prague, Czech Republic, 2011.

[HÉN 12] HÉNO R., CHANDELIER L., SCHELSTRAETE D., "Costless platform for high resolution stereoscopic images of a high gothic façade", *Proceedings of the ISPRS symposium*, Melbourne, 2012.

[HIR 08] HIRSCHMÜLLER H., "Stereo processing by semiglobal matching and mutual information", *IEEE Transactions on Pattern Analysis and Machine Intelligence,* vol. 30, no. 2, pp. 328–341, 2008.

[HOT 75] HOTTIER P., "Photogrammétrie analytique par arcs de courbe homologues ; application au chantier de Massy-Vilaine, Etude faite pour le compte du Centre d'Etudes des Bâtiments et des Travaux Publics", Report of the National Geographic Institute, 1975.

[ICO 72] ICOMOS, Photogrammétrie des Monuments et des Sites, Saint-Mandé, France, 1972.

[IGN 48] IGN, "La stéréophotogrammétrie au service de l'architecture", *Actes du Congrès International de Société Internationale de Photogrammétrie*, Commission 5, 1948.

[IGN 67] IGN, Institut Géographique National, Photogrammétrie Architecturale – Application des Méthodes Photogrammétriques au Relevé des Monuments, *brochure*, Paris, p. 36, 1967.

[IGN 92] IGN, GEOFIL, Magazine de l'Institut Géographique National, vol. 9, pp. 1–5, 1992.

[KAS 01] KASSER M., EGELS Y., *Photogrammétrie Numérique,* Hermès Science, Lavoisier, Paris, p. 378, 2001.

[KUT 00] KUTULAKOS K., SEITZ. S., "A theory of shape by space carving", *International Journal of Computer Vision,* vol. 38, no. 3, pp. 199–218, 2000.

[LAU 99] LAUSSEDAT A., *La Métrophotographie,* Edition Gauthier-Villars, Paris, p. 52, 1899.

[LEY 09] LEYMANN F., "Cloud computing: the next revolution in IT", *Proceedings of the 52nd Photogrammetric Week,* Stuttgart, Germany, 2009.

[LOW 04] LOWE D., "Distinctive image features from scale-invariant keypoints", *International Journal of Computer Vision,* vol. 60, no. 2, pp. 91–110, 2004.

[MAÎ 92] MAÎTRE H., LUO W., "Using models to improve stereo reconstruction", *Journal IEEE Transactions on Pattern Analysis and Machine Intelligence,* Special issue on Interpretation of 3D Scenes - Part II, vol. 14, no. 2, pp. 269–277, February 1992.

[PAP 12] PAPARODITIS N., PAPELARD J.-P., CANNELLE B., *et al.*, "Stereopolis II: a multi-purpose and multi-sensor 3D mobile mapping system for street visualisation and 3D metrology", *Revue Française de Photogrammétrie et de Télédétection,* vol. 200, pp. 69–79, 2012.

[PIE 06] PIERROT-DESEILLIGNY M., PAPARODITIS N., "A multiresolution and optimization-based image matching approach: an application to surface reconstruction from SPOT5-HRS stereoimagery", *ISPRS Workshop on Topographic Mapping from Space (With Special Emphasis on Small Satellites),* Ankara, Turkey, February 2006.

[PIE 11] PIERROT-DESEILLIGNY M., CLÉRY I., "Évolutions récentes en photogrammétrie et modélisation 3D par photo des milieux naturels", in JAILLET S., PLOYON E., VILLEMIN T. (eds.), *Images et Modèles 3D en Milieux naturels,* Le Bourget-du-Lac, Collection Edytem, vol. 12, pp. 51–66, 2011.

[PIE 12] PIERROT-DESEILLIGNY M., CLÉRY I., "Some possible protocols of acquisition for optimal use of the 'APERO' open source software in automatic orientation and calibration", *Tutorial for EuroCow,* Castelldefels, Spain, 8–10 February 2012.

[PIN 11] PINTE A., BAILLARD C., DENIS E., "Modélisation de façade par analyse conjointe d'images terrestres et de données laser", *Revue Française de Photogrammétrie et Télédétection,* vol. 194, pp. 53–64, May 2011.

[REM 08] REMONDINO F., EL-HAKIM S., GRUEN A., *et al.*, "Turning images into 3D models – Development and performance analysis of image matching for detailed surface reconstruction of heritage objects", *Signal Processing Magazine,* vol. 25, no. 4, pp. 55–65, 2008.

[RIN 13] RINAUDO F., CHIABRANDO F., "Calibrating and evaluating a range camera for cultural heritage metric survey", *Proceedings of the 24th CIPA Symposium,* Strasbourg, France, 2013.

[ROT 13] ROTENBERG K., SIMARD L., SIMARD P., "Dense DSM generation using the GPU", *Proceedings of the 54th Photogrammetric Week,* Stuttgart, Germany, 2013.

[ROU 36] ROUSSILHE H., *La Photogrammétrie et ses Applications Générales,* Volume 1, Principes Généraux et Restitution de Clichés Isolés, Editions Eyrolles, 1936.

[SAM 13] SAMAAN M., HÉNO R., PIERROT-DESEILLIGNY M., "Close-range photogrammetric tools for small 3D archeological objects", *Proceedings of the 24th CIPA Symposium*, Strasbourg, France, 2013.

[SCH 02] SCHARSTEIN D., SZELISKI R., "A taxonomy and evaluation of dense two-frame stereo correspondence algorithms", *International Journal of Computer Vision,* vol. 47, nos. 1–3 , pp. 7–42, 2002.

[SEI 06] SEITZ S., CURLESS B., DIEBEL J., *et al.*, "A comparison and evaluation of multi-view stereo reconstruction algorithms", *Computer Vision and Pattern Recognition,* IEEE Computer Society Conference, vol. 1, pp. 519–528, 2006.

[SNA 06] SNAVELY N., SEITZ S., SZELISKI R., "Photo tourism: exploring photo collections in 3D", *ACM Transactions on Graphics (SIGGRAPH Proceedings),* vol. 25, no. 3, pp. 835-846, 2006.

[SYL 13] SYLAIOU S., BASIOUKA S., PATIAS P., *et al.*, "The volunteered geographic information in archaeology", *Proceedings of the 24th CIPA Symposium*, Strasbourg, France, 2013.

[TIA 10] TIAN Y., GERKE M., VOSSELMAN G., *et al.*, "Knowledge-based building reconstruction from terrestrial video sequences", *Journal of Photogrammetry and Remote Sensing*, vol. 65, no. 4, pp. 395–408, July 2010.

[VAL 11] VALLET B., PIERROT-DESEILLIGNY M., BRÉDIF M., *et al.*, "Amélioration d'une base de données d'empreintes de bâtiments pour la reconstruction 3D: une approche par découpe et fusion", *Revue Française de Photogrammétrie et de Télédétection,* vol. 195, pp. 42–50, 2011.

[VIN 13] VINCENT R., POIVILLIERS J., "Georges Poivilliers 1892–1968", *REVUE XYZ*, vol. 134, pp. 45–60, 2013.

[VU 09] VU H.H., KERIVEN R., LABATUT P., *et al.*, "Towards high-resolution large-scale multi-view stereo", *Conference on Computer Vision and Pattern Recognition*, Miami, June 2009.

[WAL 94] WALDHÄUSL P., OGLEBY C., "3x3-Rules for simple photogrammetric documentation of architecture", *International Archives of Photogrammetry and Remote Sensing*, vol. 30 (part 5), pp. 426–429, 1994.

[ZHA 11a] ZHANG Y., XIONG J., HAO L., "Photogrammetric processing of low-altitude images acquired by unpiloted aerial vehicles", *The Photogrammetric Record,* vol. 26, no. 134, pp. 190–211, June 2011.

[ZHA 11b] ZHANG Y., HU B., ZHANG J., "Relative orientation based on multi-features", *ISPRS Journal of Photogrammetry and Remote Sensing,* vol. 66, no. 5, pp. 700–707, September 2011.

3

3D Digitization by Laser Scanner

The technique of digitization by laser scanner is relatively new, especially when compared to the age of the photogrammetric surveying methods discussed in the previous chapter. Nevertheless, in a short period, lasergrammetry has become an effective tool, adapted to the 3D recording of a site, especially outstanding structures. This chapter examines the instruments required for this type of survey, and then describes the methods to implement in order to acquire and process data from laser scanners with a view to constructing 3D models in the form of point clouds.

3.1. History

Laser emission was carried out for the first time in 1960 by Theodore Maiman [MAI 60]. Its use in remote sensing soon followed with the measuring of the distance from the earth to the moon two years later [SMU 62] using the principle of lidar technology (*light detection and ranging*), which relies on the analysis of properties of a laser beam sent toward an emitter. The industrialization of this technique took off in 1964 with the Geodolite[1,2], produced by

---

1 http://amhistory.si.edu/surveying/object.cfm?recordnumber=748473.
2 http://historywired.si.edu/object.cfm?ID=22.

Spectra-Physics (Mountain View, California), which enabled the measurement of profiles by mounting the instrument on an aircraft attachment.

The technology gradually developed to all observation scales (equipment mounted on satellites or airplanes, or used from the ground) with evolutions mainly regarding acquisition frequency, the possibility of scanning surfaces and the increased instrumental range. At the end of the 1990s, several manufacturers were offering terrestrial laser scanners with a measurement rate of around 5,000 points per second. This improved performance enabled new applications in the field of topography, especially for carrying out high-resolution 3D surveys of works of art or buildings. The first tests in the field of heritage conservation took place around the same time: firstly with instruments designed in research laboratories [WEH 99, FAN 99] then with commercial laser scanners[3] [BAR 01, BOE 01, GOS 03, WIL 03].

Since then, the technology has improved further, leading to more and more compact instruments, able to capture entire scenes in the form of millions of 3D points, in a very short time. Larger sites can now be captured in 3D. Today, the equipment is widely available and is an important feature in the surveyor's tool box. Large-scale heritage digitization projects have been carried out on all five continents[4].

Due to their effectiveness, terrestrial laser scanners are essential tools for digitizing complex areas, including outstanding sites which, due to their architectural boldness, are difficult to reconstruct with traditional methods such as tacheometry and photogrammetry.

---

3 http://www.ensg.eu/La-cooperation-du-DIAS-avec-le-Yemen.

4 http://archive.cyark.org/.

3.2. Equipment

3.2.1. *How a terrestrial laser scanner works*

A terrestrial laser scanner or 3D scanner is an instrument, which enables the 3D digitization of an area by laser scanning. The main components of this type of equipment are:

– a laser emitter generating a pulse at the desired frequency and power (Figure 3.1(1));

– a scanning system using a rotating mirror or a prism which deflects the beam in order to move the position of the laser pulse on the object of interest. Evidently, it is necessary to know the vertical and horizontal angles of the output beam by angular encoders (Figure 3.1(2));

– an optical receiver system recording the return of the beam (Figure 3.1(3));

– operating software enabling the setting of scan parameters and the necessary pre-processing to obtain usable data (Figure 3.1(4)).

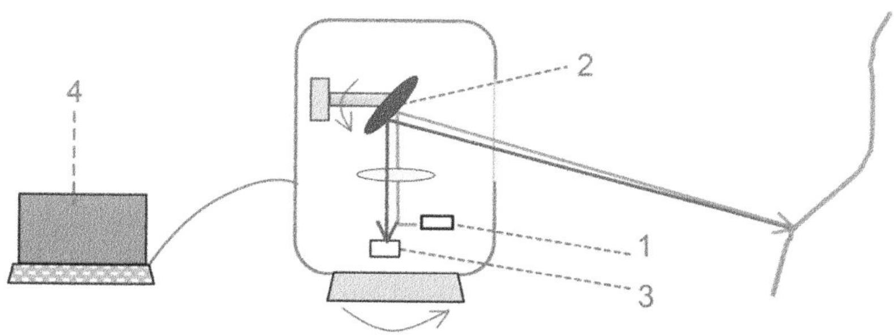

Figure 3.1. *Simplified diagram of a laser scanner*

The data obtained are in the form of 3D point clouds in the instrument coordinate system, initially spherical coordinates (ρ, ϕ, θ) with ρ being the distance from the center. These values are generally converted into Cartesian coordinates (X, Y, Z). 3D points acquired in this way are often accompanied by an intensity value (strength of the returning laser signal) and, if a camera is present, by color values in red, green and blue channels.

3.2.2. *Various families of laser technology*

There are a number of different types of technology for acquisition by laser scanning, the principle of which is concisely summarized in this section. For a more detailed description of how the instruments work, refer to [LER 08, BER 10] and [LAN 11a].

– *Time-of-flight or pulse laser scanner*

The instrument measures the time taken by the laser beam between its emission and its return after its reflection by the object with which it made contact. The distance between the laser and the scene can be directly calculated from this. To detect the reflected beam, various criteria are used by the instruments including the extraction of the peak of the returning signal, thresholding of the returned intensity and a calculation of the amplitude ratio of the beam.

– *Phase shift laser scanner*

The laser is emitted continuously, at high intensity with sinusoidal amplitude modulation. The distance between the laser and the scene is determined by comparing the phases of the wave emitted and the wave that is reflected. Other scanners, which are less adapted to ground-based surveys because of a lower measurement rate carry out a modulation on the frequency of the wave.

– *Triangulation-based laser scanner*

The returning beam is observed from an optical receiver system, which is placed at a given distance from the source of the laser emission. The position of the object of interest is determined by the resolution of the triangle. In the past, this type of scanner was used for surveys on parts of a building. Nowadays, this technique is mainly used with hand-held scanners, the principle of which is similar to that of triangulation scanners, apart from the fact that the laser emitting source is mobile, and it is therefore necessary to know its position at all times. They are used to digitize small objects with high levels of accuracy.

These three types of technology fundamentally differ in the way they determine (function f in Figure 3.2) the distance between the scanner and the object to be digitized. Some manufacturers offer laser scanners which combine these various methods in order to improve the performance of the instrument[5]. The following section will examine the instrumental characteristics of each of these types of technologies. Examples of the equipment used are shown in Figure 3.3.

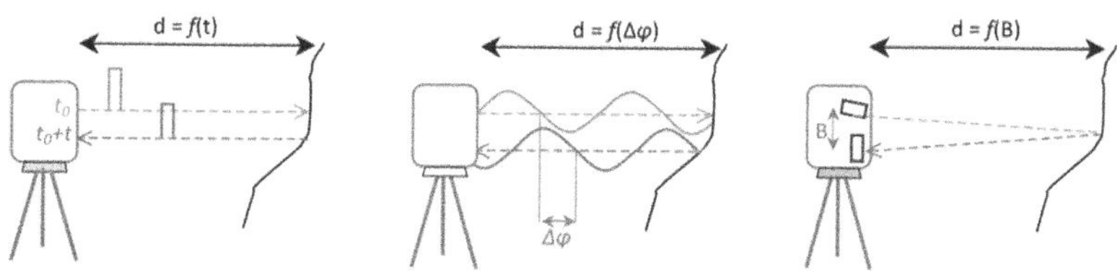

Figure 3.2. *How the three main types of laser scanner work: time-of-flight (left), phase shift (center) and triangulation (right)*

5 http://www.trimble.com/3d-laser-scanning/cx.aspx?dtID=overview&.

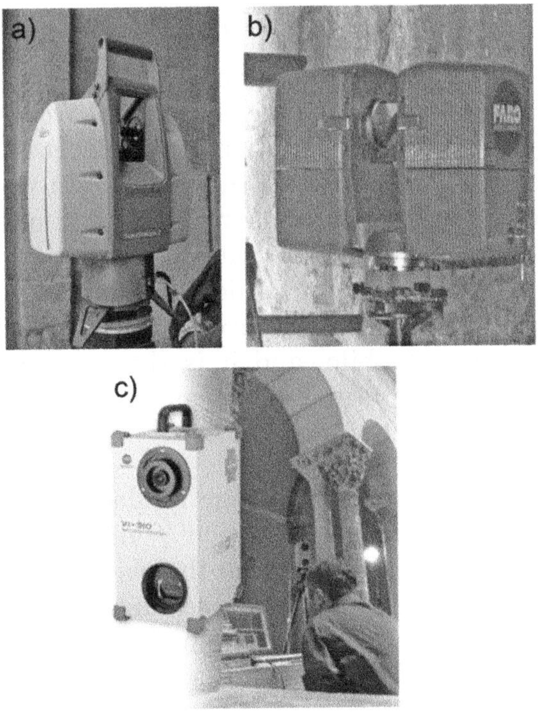

Figure 3.3. *Examples of laser scanners: a) time-of-flight, b) phase shift and c) triangulation*

Moreover, to meet the equivalent requirements to those discussed in section 2.3.4.3 regarding the mobile acquisition of images, there are systems where one or more laser scanners are mounted on mobile vehicles, either on their own, or accompanied by cameras. These systems can be ground-based [PAP 12] or airborne (airplane, glider, helicopter, UAV). There are many applications for this technology, the main ones being the 3D digitization of streets and the calculation of digital elevation models of large areas. Laser scanners used in this way must have a long range and be accompanied by an inertial navigation system and a GNSS antenna which are able to record the position and orientation of the moving instrument at all times.

Finally, although it is not the subject of this text, other types of measurement technology enable the production of 3D surveys (structured-light or modulated-light scanners), which are more usually used in the industrial field, for small areas.

3.2.3. *Characteristics of terrestrial laser scanners*

There are many parameters that characterize terrestrial laser scanners. These parameters are listed below, and typical values for each technology are summarized in Table 3.1:

– measurement rate (points per second): the highest rates currently exceed one million points per second and can be obtained by phase shift scanners;

– minimum range (meters): terrestrial laser scanners can only take measurement starting from a given value from several decimeters to a meter, variable according to the technology used;

– maximum range (meters): the laser beam is emitted at a given power which, due to the attenuation that occurs during the path of the beam, limits the acquisition to a specific area. The depth of this area depends on the power of the laser and the reflectance of the object (fraction of light energy reflected dependent on its wavelength and the angle of incidence of the laser). Manufacturers, therefore, specify the instrumental range for a given reflectance of the object (usually 90%). For phase shift scanners, range is also limited by the technology, due to the integer cycle ambiguity linked to the introduction of a new cycle in the modulation of the wave. As a result, maximum distance is extremely variable – from a meter to several kilometers – according to the technology of the laser being used;

– horizontal and vertical resolution or incremental motor (degree or millimeter): this is the minimum angle between two adjacent positions of the laser; horizontal and vertical. It is possible to express this as a minimum distance between two adjacent points on the object at a given distance. It is clearly possible to adjust acquisition to higher values, and the term "acquisition step size" can therefore be used.

The combination of horizontal and vertical step sizes helps to control density (in terms of points per meter) on the object, even if this density is irregular in practice.

– beam diameter at exit and divergence (millimeters and radians): when it leaves the instrument, the laser beam is of a certain size and forms a given angle;

– range accuracy (meters): this parameter allows the user to assess the true value of the distance given by the laser measurement system. As it reduces with distance, accuracy is generally given to within one sigma and for a given distance;

– angular accuracy (degrees): this parameter assesses the angular position given by the angular encoders.

The positional accuracy of any point measured on the object is, therefore, a combination of these last two parameters, to which it is necessary to add the nature of the interaction between the laser beam and the object. Indeed, the granularity of the surface, its reflectance and the angle of incidence all have an influence on the way that the beam is reflected. This topic, amongst others, is discussed in section 3.2.4.

Furthermore, it should be noted that the resolutions suggested by many instruments allow a more refined scanning than the accuracy it is able to achieve. Therefore, the result is sometimes a dense but noised point cloud.

– horizontal and vertical fields of view (degrees): these parameters indicate the area visible by the instrument, which is limited by material constraints;

– wavelength (nanometer): the laser beam corresponds to one monochromatic wave whose wavelength, for terrestrial laser scanners, is between 530 and 1550 nm, from green to near-infrared;

– safety classes: the manufacturer must indicate the safety class of the laser scanner, described by the International Electrotechnical Commission (IEC) standard 60825-1 ed2.0 [IEC 07] as set out by the IEC, which categorizes according to four main laser danger classes. The laser scanner safety classes normally used in the context of topographical ground-based surveys range from *1* to *3R* and *3B*. The latter classes more specifically include safety measures that must be respected.

As well as these parameters, which can be found on any type of laser scanner, other optional aspects may also be given:

– verticalization system: some laser scanners are able to align the instrument's vertical with the local vertical, using an integrated level compensator;

– centering system: some laser scanners enable the centering of the scanner on a given point due to an optical or laser plummet and a mark for measuring the instrument height;

– camera: a photographic or video camera is occasionally included in order to obtain the color of the object at the point measured. To complement the volume obtained and the intensity of the reflected signal, this information can be useful in a number of ways: presence of paintings, requirement to assess the state of the material, to produce a realistic rendering, etc. Due to the mechanical difficulty of aligning instruments' rotational axes, a geometric shift is often observed between the camera and the laser even after correction, which can be carried out in pre-processing. As a result, the use of this information is often limited. Moreover, the cameras currently used provide images of a significantly lower quality than those acquired by the single-lens reflex cameras that are available. Consequently, it is often a requirement nowadays that complementary image surveying be carried out by external photographic equipment (see section 3.3.5);

– full-wave form scanner: this type of equipment allows the processing of the whole returned signal, which can give information on both the material with which contact has been made (especially its rugosity, by studying the form of the returned wave) as well as all objects encountered in the path of the laser beam (with an ability to discriminate between objects in terms of depth). This final point is particularly interesting in the case of obstructions (e.g. tree branches) which are in the path of the laser beam [DON 09].

Technology / *Parameter*	*Time-of-flight*	*Phase shift*	*Triangulation*
Measurement rate	10,000 to 300,000 points/s	1 million points/s	Tens of thousands of points/s
Minimum range	1–5 m	0.3–0.5 m	from 0.1 m
Maximum range	300–6000 m	80–180 m	1.0–2.5 m
Horizontal and vertical resolution	0.0002–0.0125°	0.001–0.002°	0.1–0.3 mm
Beam diameter at exit	6–18 mm	3 mm	–
Beam divergence	0.15–0.35 mrad	0.16–0.25 mrad	–
Range accuracy	3–5 mm @50 m	2–3 mm @50 m	<1/10th mm
Angular accuracy	0.0002–0.01°	0.001–0.007°	–
Horizontal and vertical fields of view	H: 360° V: 60–360°	H: 360° V: 270–360°	0.5 x 0.5 x 0.22 mm to 1.5 x 1.1 x 1.7 m
Wavelength	532–1550 nm	670–905 nm	Around 660 nm
Safety class	1 or 3B or 3R	1 or 3R	1
Weight	10–20 kg	5–15 kg	2–7 kg

Table 3.1. *Typical values for instrumental characteristics according to technology (based on[6] and [REM 11])*

6 http://www.geo-matchingmatching.com/category/id46-terrestrial-laser-scanners.html.

3.2.4. *Errors affecting measurement and calibration*

Some errors affect the measurements obtained through terrestrial laser scanners either at the level of the distance obtained and the angles (hence the spatial positioning of points measured) or returned intensity.

3.2.4.1. *Geometric aspects*

These errors can be categorized according to their origin [SCH 07, RES 09, LIT 10]:

– various errors linked to physical aspects of the laser measurement. The distance measurement is subject to a number of factors which reduce its accuracy: thermal noise, quantum noise and dark current. It is possible to limit the impact of these factors by taking several consecutive measurements and then computing the average, but this obviously slows down the acquisition speed;

– systematic errors linked to equipment and its construction:

- individual quality of each of the components of the laser scanner has a direct impact on the result: quality of angular encoders, analog-to-digital converter, timer accuracy, type and performance of oscillating mirror used, etc.;

- inaccuracies in the assembly of the various components, including non-alignment or non-orthogonality of axes, leading to errors regarding angles and distance;

- internal, optical or electric interferences can also reduce the performance of the instrument;

– factors linked to the environment:

- the wave propagation speed is dependent on the environment, characterized by its refraction index.

Variations can also arise due to the nature of the surrounding environment and the weather, because of the changes in temperature, pressure or humidity. Dry and still conditions are most suitable, and laser beams through water or glass should be avoided;

- ambient light may disturb the measurements and carrying out acquisitions at night may be recommended in some cases;

– factors caused by the interaction between the laser scanner and the object of interest:

- divergence of the laser beam means that, at the edges of the object, part of the wave may be sent back to the scanner and part may continue its path toward another object. If two objects are close together, in terms of the depth resolution of the laser scanner, the detector will not be able to distinguish between the two returned signals and the 3D point obtained will be positioned between the two objects, thus giving a mixed pixel;

- the laser beam's angle of incidence on the object has an impact on the level of quality of the signal returned [SOU 11];

- highly reflective objects may disrupt adjacent objects and will be falsely amplified in the point cloud. This can also be seen in the trails in front of and behind the object;

- specular surfaces can create multiple paths, especially at short distances;

- dark objects with weak reflectance, return little energy. The beam returned then has a worse signal-to-noise ratio, and the distance measured is consequently more sensitive to the type of return detector;

- the laser beam is ineffective on some specific materials: material penetration (marble [GOD 01], polystyrene, etc.), total reflection (gilding, dark and highly reflective stained wood), transmission through glass. In the latter case, the beam is partly returned and partly transmitted with a modified path.

In a more general sense, we should refer to [VÖE 08] to have an idea of the impact of different materials on laser measurements.

A calibration of the laser scanner can be carried out in order to determine some systematic errors, either by physical modeling if the source of the problems is known, or by empirical, often polynomial, modeling. Artifacts and random errors cannot be corrected by this type of procedure and are often processed manually.

There are a number of different calibration strategies:

– the various components of the laser scanner are calibrated separately, which helps to identify the source of the problems and requires specific equipment for individual calibrations;

– the entirety of the system is calibrated on a specific site composed of targets acquired from various positions and orientations. The unknown parameters (positions and orientations of the scanner stations, positions of targets and parameters of the calibration model) are determined using the redundant information produced by this survey. This method is similar to the process used to calibrate the camera's internal parameters. Particular attention must be paid to the distribution of observed targets (in particular, to ensure different distances, in the whole of the scanner's field of view) and the production of redundant observations in order to dissipate the ambiguities due to the correlation between the various parameters to be determined;

– another solution is to use characteristic objects, such as planes, during a calibration process, by on-site measurements, thus simplifying procedures but not necessarily ensuring an equivalent accuracy.

The specifications ultimately provided with the equipment, especially the accuracies indicated (see Table 3.1), take into account the residual errors, which were not able to be corrected during the calibration process. This is usually carried out by the manufacturer.

3.2.4.2. *Radiometric aspects*

The intensity of the returned laser signal is a measure, which can contribute to an understanding of the scene by giving information on the materials of which the objects are made. For some applications, it may be necessary to have an absolute measure. Besides the influence of the object of interest's reflectance, which is being looked for in practice, the intensity is modified by a number of other factors:

– the energy decreases inversely to the square of the distance;

– the atmosphere attenuates the signal as well as the components of the scanner which transmit or reflect the laser beam;

– the signal is amplified during the analog-to-digital conversion;

– some pre-processing that is carried out, the nature of which it is difficult to ascertain.

Eventually, tests [PFE 07, KAA 09, KRO 13] aiming to evaluate the quality of the intensity measured by the laser scanners show that it is difficult to establish common rules to all scanners, especially because radiometric calibration is carried out in different ways depending on the manufacturer. Some software for pre-processing point clouds do not correct these errors, others include a correction, the content of which is unknown to the user.

To the best of our knowledge, there are no evaluations on the quality of digital data from the cameras which accompany the laser scanners.

3.2.5. *The main manufacturers*

For time-of-flight or phase shift laser scanners, the market consists of around ten different manufacturers. These companies are often in direct competition; some offer products which are very similar in terms of performance. The website geo-matching.com offers a regularly updated comparison of the instruments which are available.

For triangulation scanners, the market is significantly different, since this process involves highly refined digitization of small objects, which is more closely linked to the world of CAD (computer-aided design). The cost of this equipment is generally lower, ranging from a few hundred to several thousand euros, depending on the required accuracy of the reconstruction.

The future appears to be moving toward multi-instrumental platforms combining accurate topometric measurements, high-density digitization by laser scanner (1,000 points/s at up to 300 m), GNSS antenna and high-quality photographic equipment in order to texture point clouds[7].

3.2.6. *Instances of use*

In the case of the digitization of an outstanding site, the three technologies described can be considered complementary:

---

7 http://www.leica-geosystems.com/en/Leica-Nova-MS50_103592.html.

– the range of time-of-flight laser scanners enables the user to target distant parts of the structure, which is often very useful, either because the building is particularly large or because greater distance between the equipment and the structure would give a better, more perpendicular view-point;

– the performance of time-of-flight scanners in natural light means that they are often preferred for outdoor acquisitions;

– phase shift scanners are particularly suited to indoor acquisitions, especially for small rooms or areas of organizational complexity, such as stairs, due to their speed and the scope of their field of view;

– the triangulation scanners can be used if the monument has architectural, decorative or furniture elements, which require more refined digitization in order to comply with the demands to the building specifications.

Work carried out on the Charterhouse of Villeneuve-lès-Avignon demonstrates this complementarity [FAV 10]: time-of-flight and phase shift scanners were used to digitize the entirety of the site, and a triangulation laser scanner was used for the acquisition of smaller or more detailed objects. The example of Cluny Abbey [LAN 11b] also shows a joint use of the various pieces of equipment adapted to the object of interest to be surveyed.

Finally, it is necessary to note the effectiveness and reliability of laser measurements as a whole, for almost all materials apart from the cases mentioned in section 3.2.4.

3.3. Acquisition by lasergrammetry

Terrestrial laser scanners are relatively easy instruments to implement. It is possible to learn the necessary expertise to set up the equipment and lead the digitization process after just a few days of training. The survey design and the creation of a network of control points, along with a hint of creativity, are where the real skill lies. Indeed, this helps to avoid errors, which can make using the survey impossible if they are discovered too late, at the post-processing stage, when access to the site is no longer available.

This is even more relevant for complex structures in which there can be many potential sources of error. This section, therefore, describes the aspects to take into account during the lasergrammetry acquisition process and especially those involved in the processing of this type of structure.

3.3.1. *Survey design*

Reconnaissance is carried out by complying with the recommendations provided in section 1.4.1. Other specific elements linked to laser acquisition must also be considered. Indeed, the spherical geometry of acquisition and the limits of the equipment's field of view mean that several different laser stations are usually required to cover the whole of the area to be digitized. It is a common presumption that the number of stations should be kept to a minimum in order to save time and simplify the acquisition. In practice, there are cases where looking for the optimum set-up can introduce undesirable complexities: non-perpendicular viewpoint, less effective control point positioning, risky station set-up or even data voids if the plan is "too tight". It is, therefore, advised to have a level of flexibility especially for high-rate acquisition systems and those whose stations are easy to set up. If laser scanners with specifically different characteristics are available, especially in terms of range and field of view, it is clearly recommended that a given scanner is used over another by virtue of its relevance for a given project.

The overall number of stations may be very high: on the Amiens Cathedral, for a complete acquisition of the structure of the monument (interior, exterior and at all levels) in 1 cm resolution, by extrapolating the figures from the first acquisition campaigns [HÉN 11], the order of magnitude is of around 100 laser stations for around 2 billion points. Evidently, this number depends on the size of the monument but also, and most importantly, on its complexity. Large rooms with relatively simple architecture often require considerably fewer laser stations than a labyrinth of interlocking corridors and stairs or sculpted façades.

3.3.2. *Requirements for control points*

A terrestrial laser scanner provides coordinates of the points measured in a coordinate system linked to the instrument. Two stations in two different places, therefore, have two independent coordinate systems that need to be connected: this is known as cloud alignment.

This relative transformation can be done by using tie points (shared regular points) or overlapping sections (shared areas between the point clouds of the different laser stations).

In order to use data in a reference coordinate system, positioning and orienting the survey in an absolute system, it is also necessary to proceed to the georeferencing of the laser stations by using ground control points.

As stated in section 1.4.1, points of different types can be used to achieve successful operations: natural details, artificial targets and geodetic points. Nevertheless, some recommendations can be given regarding the nature of the laser measurement and the acquisition configurations. Dark materials reflect little, or no signal at all and should, therefore, be avoided. Geometric shapes which do not change depending on the point of view, such as a sphere, are well-suited even though they can be less discrete and more complicated to position on the site. Correctly defined and well-maintained stone edges could also be used as highly visible decorative elements. The choice of natural points used should be adapted to the scanner being used especially in accordance with the quality of the intensity obtained. Moreover, a specific acquisition on a very small resolution, around a few millimeters, is generally carried out on these details to be able to measure them as accurately as possible in the laser point cloud. Figure 3.4 shows examples of appropriate control points for laser scanners.

Figure 3.4. *Suitable control points for laser digitization*

In the case of a station for which the scanner has been verticalized and possibly centered on a point, this provides further complementary knowledge on registration (local vertical and center of the laser coordinate system). This can allow for a reduction in the number of control points necessary.

Regarding the distribution of points around the site, the following rules should be followed:

– if possible, the elements should be distributed throughout the whole of the scanner's field of view, at different distances and, where possible, far from the station;

– it is preferable to obtain multiple views of the points, since this helps to consolidate the network and improve the level of control;

– configurations which do not fix all the degrees of freedom (aligned points, areas of plane overlap) should be avoided.

The complexity caused by outstanding sites often limits the ability to establish a network of control points. The specificities of each site need to be taken into account for this process. The aim at this point is not to make exhaustive recommendations (which would be impossible given the number of cases with which one may be confronted), but rather to make the coordinator of a lasergrammetry project aware of the importance of the design of such a network. Section 3.4 discusses the geometric accuracy obtained according to the nature of the network and the variables which can allow this network to be designed as best as possible.

The following examples highlight the difficulties that may be encountered:

– natural alignment of the elements to be digitized such as a corridor without openings or a succession of rooms: it is, therefore, advisable to use areas of overlap between point clouds to minimize ambiguities, especially by covering the ends and the edges (Figure 3.5). Control points should be distributed at the maximum on all sides of the alignment formed by the laser stations, horizontally and vertically.

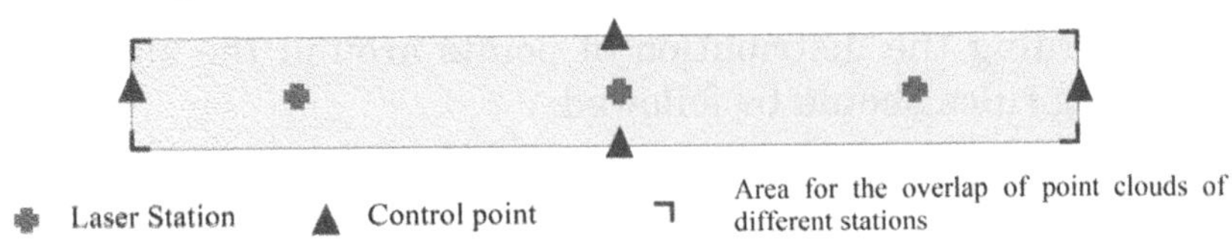

Figure 3.5. *Example of equipment with control points suitable for a long narrow corridor*

– for areas which are difficult to access, such as wells or underground rooms, complementary devices can be used. For instance, the laser scanner can be placed on specially adapted supports or be carried by specific equipment (Figure 3.6);

Figure 3.6. *Examples of ways to adapt to a study site: digitization in a cistern [BOR 10] and in a well*

– interior and exterior: although this is not specific to complex buildings, particular attention should be given to this aspect. Indeed, it is important to create multiple links between interior and exterior (by openings created by doors and windows) in order to avoid deformations caused by a heterogeneous distribution of points. These links can be ensured by control points (Figure 3.7) or by placing the laser stations directly in the openings themselves.

Figure 3.7. *Positioning of a sphere in an opening*

3.3.3. *Practical realization*

The constraints discussed previously can be very high or even contradictory; therefore, it is sometimes necessary to find a compromise between the various requirements given. In the most complex cases, the pragmatism and ingenuity of survey designers can be put to the test. Experience will of course help in making the right choices. In the case of laser use, other issues need to be considered to heed the warnings for use contained in the safety class given. Night work may be necessary as well as closing off access for part of the site.

In all cases, the implementation of a laser station requires the realization of well-established tasks, which can be included on a checklist, in order to ensure that nothing is omitted. Moreover, it is recommended that particular attention is given to the station's documentation (paper or digital), which often enables the correction of errors made and to highlight ambiguities.

Figure 3.8. *Aerial view of part of the lasergrammetry site at St. Symeon (Syria): in total, the site covers more than 5 km² and has 297 laser stations (black points on the map)*

3.3.4. *Advantages and limits of this technology*

Terrestrial laser scanners enable the user to directly obtain comprehensively accurate and reliable 3D point clouds. They are, therefore, highly effective digitization instruments, suited to the surveying of large architectural areas.

Their limits are as follows:

– the cost of the equipment remains high, even though the market nowadays is offering equipment at lower and lower prices;

– the cumbersome nature of operation, even though the latest models are more compact, lighter and contain an onboard controller which makes them easier to use. However, outstanding sites often have sculpted decorative parts, which are easier to digitize if the user can move freely with the measurement instrument, which is possible with hand-held scanners or photogrammetric techniques;

– inaccessible areas: this is the case for high exterior areas, which can only be reached by using light-weight carriers (glider, UAVs and poles), which can be obtained more easily with photographic equipment. Equivalent equipment for laser scanners should be available in the market soon[8]. For more large-scale applications, some companies[9] put laser scanners onto helicopters, for instance in the case of surveying corridors (roads and railways), for modeling engineering works (dams and bridges) or for geological analysis;

– unsatisfactory quality of information provided by the cameras of the current scanners, both for the color and the localization of this information;

– requirement of working on ground, which is sufficiently stable during the acquisition phase.

3.3.5. *Linking of external photographs*

For some applications, such as a virtual visit or if the site contains interesting colors, a realistic rendering of the environment may be requested. While the market waits for laser scanners which contain sufficiently effective cameras, alternative solutions have been created to provide high-quality colorimetric information:

– photographs can be taken by independent equipment. The rules for obtaining detailed and high-quality images outlined in Chapter 2 are also valid here. However, the geometric acquisition constraints are quite different, since stereoscopy is not essential. For several years, a large amount of research has been carried out on automatic

8 http://yellowscan.lavionjaune.com/.
9 http://www.helimap.ch.

georeferencing with laser point clouds in order to process this complementary data as quickly as possible (see section 5.3.5);

– to obtain high-quality images in the acquisition geometry of the laser scanner, it is possible to use a panoramic acquisition head holding a digital camera (Figure 3.9) positioned as close as possible to the center of the laser scanner, if necessary by adding machined parts to ensure this connection or by placing the camera on the laser scanner. Processing is necessary in order to accurately combine the geometry of the two types of data [MEI 10, CRA 12].

Figure 3.9. *Acquisition system for panoramic images (left) and an example of a panoramic image (right)*

A consideration should be made as to the choice of the positioning of the image stations. Indeed, the placement of laser stations is highly constrained by the realization requirements of the site to be digitized. As a result, the positions of the laser stations are often not ideal for photographic acquisition (images with low resolution and high incidence), or for the time that they are taken (influence of lighting conditions on the quality of the images).

3.4. Registration of laser stations

In this text, the term "registration" is used in a general sense, referring to cloud alignment methods, georeferencing and their combinations. In all cases, it determines the process which allows the transformation from an instrument's coordinate system, linked to the laser scanner, to a global coordinate system, which can be linked to the laser acquisition project, the monument or be a reference system. This transformation consists of a rotation $R_{Instument \to Global}$ and a translation $T_{Instrument \to Global}$ and has six parameters, which can be calculated using points ("point-to-point"' registration) or areas of overlap on the point clouds ("cloud-to-cloud" registration).

3.4.1. *Principle of the different registration strategies*

To determine the transformation parameters for a given station, only using a single point-to-point registration method, it is necessary to know at least the coordinates of three non-aligned points in the instrument system $M_{Instrument}$ and in the global system M_{Global}. The mathematical matrix formula is:

$$M_{Global} = R_{Instrument \to Global} \times M_{Instrument} + T_{Instrument \to Global}$$

The use of two points leaves a degree of freedom around the line which connects them. The system of equations can be resolved by using a least square adjustment which requires a system of linear equations. It is, therefore, necessary to linearize the system around an approximate solution. In practice, it is advisable to have at least five points per station to be able to detect and correct any errors. The user should attempt to distribute the points evenly between the different stations for reasons of cost-effectiveness and to strengthen the network's robustness. If the laser scanner is verticalized or even centered on one point, the number of exterior points necessary for registration is lowered: at least two points in the first case and one in the second, to which it is always necessary to add extra points, in order to detect errors.

When it is only a cloud-to-cloud registration method which is used, the output system can only be a local system linked to the acquisition or monument, and it is important to ensure that the point clouds from different laser stations have enough overlap. It should also be checked that the stations can be completed one after the other (no independent sub-areas). In this case, the calculation of the six transformation parameters is done with the methods whereby the distance between the point clouds is minimized, the most well-known and widespread of which is the ICP method (Iterative Closest Point) [BES 92], which consists of many different variants, developed to improve the initial algorithm and adapt it to the specificities of the data [POM 13]. For a review of the different methods available, refer to [RUS 01]. In all cases, the problem is still non-linear, requiring an initial solution. Consequently, it is still necessary to have tie points between the clouds, but the constraints regarding how many and how accurate they are can be relaxed. Moreover, research is currently underway to automate this part of the process as well. Finally, for the problem to be correctly determined, as stated previously, areas of overlap must not be ambiguous. In particular, the shared area must not be of constant curvature.

Logically, if both types of registration are used simultaneously, the requirement in terms of equipment is a combination of the above. There are few practical studies on optimizing the network of points in this case. It is useful to refer to the equipment in terms of control points for the realization of aerotriangulation (see section 2.4) which has been studied in depth and which is significantly simpler because of the regularity of the aerial image survey and the possibility offered by photogrammetry software to automatically extract many tie points between the images. As a minimum, three control points are needed to transfer the whole site toward the reference system but, if only three are used, local derivations are possible, and it would not be possible to detect them.

The main strategies that may be used are summarized in the Table below.

Name	Output coordinate system	Method description	Advantages	Disadvantages
Cloud alignment only	System linked to laser acquisition or monument	Various laser stations are fixed by cloud-to-cloud and/or (tie) point-to-point methods	Light-weight method in the field	Possibility of derivations Registration only relative
Georeferencing only	Reference system	Each laser station is georeferenced by a point-to-point method which only uses control points (five points per station recommended)	Little risk of derivation in relation to the ground	Non-minimized gaps between clouds Cumbersome equipment needed for control points
Cloud alignment and georeferencing	Reference system	Tie (point-to-point and cloud-to-cloud) and control points are used simultaneously	Hybrid method which reduces the requirement for control points Compromise on the constraints between relative and absolute registration	Difficult to clearly specify the requirement in terms of number of tie and control points

Table 3.2. *Description of different registration strategies for laser stations*

When a large site needs to be processed, it is advisable to work with sub-areas and gradually add laser stations to be registered. This allows for easier detection of errors and gradual correction as well as verifying the quality of the georeferencing sub-area by sub-area. Overall adjustment using all observations (tie points, control points and overlapping areas) is advised at the end of the process, but it can be challenging to calculate in extreme cases for very large sites. In this case, local calculations can be kept, which are still expressed in the global system due to the densification of the control point network. The combination of registration elements of various types, which often have variable accuracy, also requires a prior weighting of measurements according to their accuracy.

3.4.2. *Automation of all or part of the registration phase*

To improve the cost-effectiveness of the lasergrammetry project and improve the results, it is advisable to try to automate the laser station registration phase.

3.4.2.1. *Automatic extraction of artificial targets*

Regarding artificial elements, software generally includes algorithms which, with the knowledge of the pattern or shape to measure, are able to offer automatic point positioning [KRE 13]. However, it is generally necessary to have an approximate position, often provided by the selection by the operator on a point cloud close to the area of interest. Although human intervention is still necessary and the time saved in the processing phase is therefore limited, it should be noted nonetheless that the positioning of the point in the cloud is improved.

In the case of spheres, a traditional least square adjustment enables the calculation of the best sphere (whose radius is possibly known, thus making the operation more effective) corresponding to the points surveyed.

3.4.2.2. *Automatic registration by the matching of geometric primitives*

Some methods aim to build connections via geometric invariants which can be easily retrieved from one station to another [BAR 08] and [ALD 13]. They are especially used in the cases of industrial sites, which are often structured in simple geometric shapes, such as planes and cylinders. This type of strategy is sometimes used to replace manually determined initial solutions. Geometric objects are generated with a method similar to those discussed in section 5.3.4 and is complemented by a phase of matching these primitives.

Approaches which aim to use geometric primitives from images of reflectance [WAN 08] and [KAN 09] have also been successfully implemented.

This method is, however, not yet available in commercial software.

3.4.2.3. *Automatic registration with the use of external images*

Another method, when the laser scanner is fitted with photographic equipment, consists of calculating the relative position of images produced by the camera with traditional image-based methods and to deduce the position of the laser scanners, possibly refining this initial estimate with the ICP method.

In all cases, the absolute positioning will need to be determined by knowledge of ground control points in the laser point clouds.

3.4.3. *Qualification of the registration of laser stations*

Following the transformation calculations for the various laser station systems toward the site's global coordinate system, the result of the registration must be qualified to ensure that the geometric quality of the site is respected regarding the building specifications. The quality of this registration is dependent on a number of factors:

– accuracy of the laser equipment (in terms of distance and angle, see section 3.2.3);

– weather: the meteorological conditions can have an influence on the quality of the results. For instance, in high temperatures, the instrument and the equipment (especially the tripod) can undergo deformations, which has a negative effect on the geometric accuracy; rain also disturbs lasergrammetric measurements;

– environment: some sites need specific precautions to ensure a level of stability (non-sturdy flooring and movable floors);

– type, number, distribution and pointing method of control points;

– position of laser stations in regard to the object and control points network;

– accuracy of coordinates for control points in the reference system;

– methods of calculation, weighting, system of equations.

The quality of the result can be evaluated by studying the residuals of the various registration elements. These can be divided into two types:

– residuals on linking constraints which give information on relative accuracy:

- for areas of overlap between clouds, the value usually verified is the average of the distances calculated. For laser scanners generally used in the architectural field, during only the alignment phase (no constraints for control points), residuals whose order of magnitude is around 1–2 mm can be hoped for. Some software is able to assess quality by visualizing distances between clouds after calculation, which it does by coloring the points of one cloud with the distance of the gaps;

- for tie points, the residual is the distance between the coordinates calculated after transformation to the global system and the adjusted ones. The value of the residual obtained depends on a combination of several factors, the main one being how well-defined the point appears;

– residuals on controls which give information on absolute accuracy:

- for control points: the residual is the distance between the position in the reference system created by external methods and that created by calculations. Its value depends on both the accuracy of the external measurements and on the positioning of the detail in the point cloud;

- for centering: the residual is the distance between the position in the reference system and that in the center of the laser station after calculation.

Strictly speaking, accuracy values should be determined from check points – those points which are not used in the registration calculation – thus enabling an objective qualification of the result.

In practice, with the most commonly used equipment and providing that the recommendations given here have been followed, an accuracy of around 0.5–1 cm should be achieved without difficulty for relative positioning of a complete digitization of a structure. Absolute accuracy will of course depend on the positioning methods used. For example, for GNSS systems, the choice of positioning in real time or static mode will have a clear effect on the absolute accuracy.

3.4.4. *Exporting registered laser point clouds*

The registration phase enables the establishment of six parameters from each of the transformations from the laser station system to the global coordinate system of the site as well as elements of quality on the accuracy of the result. At this point, two solutions are available to the user:

– the coordinates of the 3D points can be kept in the original system linked to the scanner and transformed in real time by the software which displays or processes the data;

– the coordinates of the 3D points are projected once and for all into the global coordinate system.

The first method has the advantage of conserving the acquisition geometry and thus keeping the topological link between the points, which allow the user to see their relative position in space and to use it for some processing. The second solution enables the use of all points simultaneously.

Finally, although the visualization of the registered 3D point clouds constitutes a first rendering, which is often judged as being spectacular (Figure 3.10), further processing is generally necessary to truly obtain the useable information (see Chapter 5).

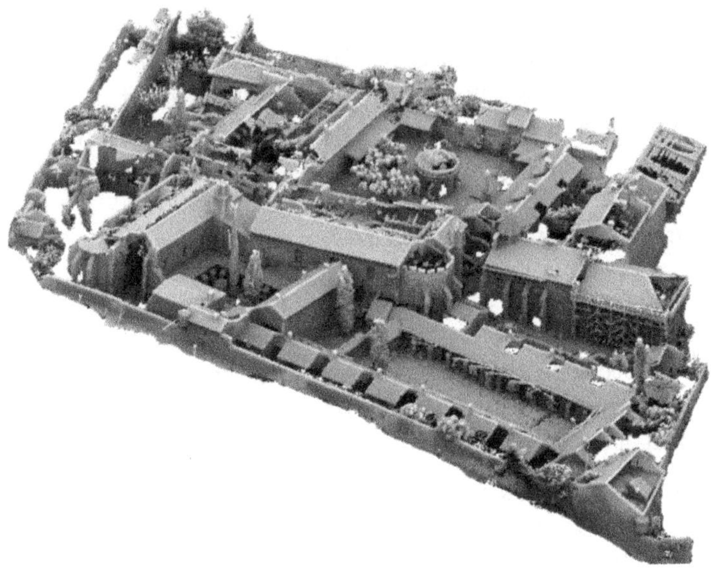

Figure 3.10. *Collection of registered point clouds of the Charterhouse at Villeneuve-lès-Avignon [GRE 09]*

3.5. Qualification of the point clouds obtained

3.5.1. *Geometric quality*

Geometric quality can be understood as the ability of the 3D model to describe the shape and position of the object of interest with regard to the specifications established.

In the case of point clouds obtained through lasergrammetry, geometric quality is mainly linked to the quality of registration (see section 3.4.3), since the accuracy of the positioning obtained by the laser scan can be, in the overwhelming majority of cases, considered as superior to that achieved by registration.

Moreover, point clouds can suffer from artifacts at a local level (see section 3.2.4):

– the noise of the distance measurement means that smooth surfaces will not be reproduced exactly as they are;

– parasite points may be present;

– random spatial sampling does not enable direct access to characteristic lines of interest and their extraction may involve approximations.

Processing can be used to (at least partially) correct these artifacts (see section 5.2.1).

3.5.2. *Radiometric quality*

The intensity of the returned laser is not widely evaluated in practice. Two explanations may be put forward:

– it is difficult to obtain absolute comparable measurements on the intensity of the laser;

– little use is made of this information as a radiometric measurement.

Furthermore, as seen in section 3.2.4, corrections of the intensity carried out by software are different from one laser scanner to another, and it is difficult to know the content of the processing carried out.

The color given by the photographic equipment connected with the laser scanner, whether internal or external, can be evaluated along the following lines:

– faithfulness of colors: same limit for the objective quality evaluation (absence of absolute information) meaning that one is often restricted to a visual estimate of the rendering;

– correct superposition of images with the laser scanner data: this can be estimated by comparing the details seen in both types of data and by studying the gap with regard to the distance from the object.

3.5.3. *Other elements of quality*

Other aspects can also be included in evaluating the quality of the point clouds obtained: completeness (missing information may remain due to disruptive elements, such as pedestrians or obstacles connected to the monument itself), reliability, interpretability and comprehension of the scene.

3.6. Bibliography

[ALD 13] AL-DURGHAM K., HABIB A., KWAK E., "RANSAC approach for automated registration of terrestrial laser scans using linear features", *ISPRS Annals of the Photogrammetry, Remote Sensing and Spatial Information Sciences*, vol. 2, part 5/W2, pp. 13–18, 2013.

[BAR 01] BARBER D., MILLS J., BRYAN P., "Laser scanning and photogrammetry: 21st century metrology", *Proceedings of the 18th CIPA Symposium*, Postdam, Germany, 2001.

[BAR 08] BARNEA S., FILIN S., "Keypoint based autonomous registration of terrestrial laser point Clouds", *ISPRS Journal of Photogrammetry and Remote Sensing*, vol. 63, no. 1, pp. 19–35, 2008.

[BER 10] BERALDIN J.-A., BLAIS F., LOHR U., "Laser scanning technology", in VOSSELMAN G., MAAS H.-G., (eds.), *Airborne and Terrestrial Laser Scanning*, Whittles Publishing, Dunbeath, Scotland, United Kingdom, p. 336, 2010.

[BES 92] BESL P.J., MCKAY N.D., "A method for registration of 3-D shapes", *IEEE Transactions on Pattern Analysis and Machine Intelligence*, vol. 14, no. 2, pp. 239–256, February 1992.

[BOE 01] BOEHLER W., HEINZ G., MARBS A., "The potential of non-contact close range laser scanners for cultural heritage recording", *Proceedings of the 18th CIPA Symposium*, Postdam, Germany, 2001.

[BOR 10] BOREL L., CABARROU M., DUBOURG S., *et al.*, "D'X,Y à X,Y,Z, de nouveaux outils pour l'étude architecturale et archéologique, Restitution 3D, lasergrammétrie et photogrammétrie: le cas de la citerne el-Nabih à Alexandrie", *Actes du colloque of Virtual Retrospect 2009, Pessac, 18–20 november 2009,* Archeovision, vol. 4, Editions Ausonius pp. 215–228, Bordeaux, 2010.

[CRA 12] CRACIUN D., PAPARODITIS N., SCHMITT F., "Image-laser fusion for in situ 3D modeling of complex environments: A 4D panoramic-driven approach", in ALEXANDRU C. (ed.), *Modeling and Simulation in Engineering,* InTech, p. 298, March 2012.

[DON 09] DONEUS M., PFENNIGBAUER M., STUDNICKA N., *et al.*, "Terrestrial waveform laser scanning for documentation of cultural heritage", *Proceedings of the 22nd CIPA Symposium*, Kyoto, Japan, 2009.

[FAN 99] FANGI G., NARDINOCCHI C., "Experiences of laser autoscanning for architecture : the domus aurea in Rome and the San Giovanni's Baptistery in Florence", *Proceedings of the 17th CIPA Symposium*, Olinda, Brazil, 1999.

[FAV 10] FAVRE-BRUN A., DE LUCA L., "De l'acquisition 3D à la réalité augmentée: le cas de l'église de la chartreuse pontificale de Villeneuve-lès-Avignon (Gard)", *Revue Française de Photogrammétrie et de Télédétection*, vol. 196, pp. 52–58, 2010.

[GOD 01] GODIN G., RIOUX M., BERALDIN J.-A., *et al.,* "An assessment of laser range measurement of marble surfaces", *Proceedings of the 5th Conference on Optical 3-D Measurement Techniques*, pp. 49–56, Vienna, Austria, 2001.

[GOS 03] GOSKAR T.A., CARTY A., CRIPPS P., *et al.*, "The Stonehenge laser show", *British Archaeology*, vol. 73, pp. 9–13, November 2003.

[GRE 09] GRESSIN A., "Conception et développement d'un système permettant l'intégration et la constitution des données hétérogènes dans un SGBD spatial", Rapport de stage du cycle ingénieur 2ème année, Ecole Nationale des Sciences Géographiques, France, 2009.

[HÉN 11] HÉNO R., CHANDELIER L., "Complementarity of terrestrial laser scanning and DSM from aerial photography: the example of the cathedral of Amiens", *Proceedings of the 23th CIPA Symposium*, Prague, Czech Republic, 2011.

[IEC 07] IEC, International Electrotechnical Commission, "Safety of laser products - Part 1: Equipment classification and requirements", International standard, IEC 60825-1 ed2.0. 30, March 2007.

[KAA 09] KAASALAINEN S., KROOKS A., KUKKO A., *et al.*, "Radiometric calibration of terrestrial laser scanners with external reference targets", *Remote Sensing*, vol. 1, no. 3, pp. 144–158, 2009.

[KAN 09] KANG Z., LI J., ZHANG L., *et al.*, "Automatic registration of terrestrial laser scanning point clouds using panoramic reflectance images", *Sensors*, vol. 9, no. 4, pp. 2621–2646, 2009.

[KRE 13] KREGAR K., GRIGILLO D., KOGOJ D., "High precision target center determination from a point cloud", *ISPRS Annals of the Photogrammetry, Remote Sensing and Spatial Information Sciences*, vol. 2, part 5/W2, pp. 139–144, 2013.

[KRO 13] KROOKS A., KAASALAINEN S., HAKALA T., *et al.*, "Correction of intensity incidence angle effect in terrestrial laser scanning", *ISPRS Annals of the Photogrammetry, Remote Sensing and Spatial Information Sciences*, vol. 2, part. 5/W2, Antalya, Turkey, 2013.

[LAN 11a] LANDES T., GRUSSENMEYER P., "Les principes fondamentaux de la lasergrammétrie terrestre: systèmes et caractéristiques", *XYZ, Revue de l'Association Française de Topographie*, vol. 128, pp. 37–49, 2011.

[LAN 11b] LANDRIEU J., PÈRE C., ROLLIER-HANSELMANN J., *et al.*, "Digital rebirth of the greatest church of Cluny Maior Ecclesia: from optronic surveys to real time use of the digital model", *Proceedings of 4th ISPRS International Workshop 3D-ARCH 2011 "3D Virtual Reconstruction and Visualization of Complex Architectures"*, vol. 38, part 5/W16, Trento, Italy, 2011.

[LER 08] LERMA GARCÍA J.L., VAN GENECHTEN B., HEINE E., *et al.*, "Theory and practice on terrestrial laser scanning", *Editorial de la Universidad Politécnica de Valencia*, 261 pages, Valencia, Spain, 2008.

[LIC 10] LICHTI D., SKALOUD J., "Registration and calibration", in VOSSELMAN G., MAAS H.-G. (eds.), *Airborne and Terrestrial Laser Scanning*, Whittles Publishing, Dunbeath, Scotland, United Kingdom, p. 336, 2010.

[MAI 60] MAIMAN T., "Stimulated optical radiation in Ruby", *Nature*, vol. 187, no. 4736, pp. 493–494, August 1960.

[MEI 10] MEIERHOLD N., SPEHR M., SCHILLING A,. *et al.*, "Automatic feature matching between digital images and 2d representations of a 3d laser scanner point cloud", *Proceedings of the ISPRS Commission V Mid-Term Symposium on Close Range Image Measurement Techniques*, vol. 38, Newcastle upon pp. 446–451, Tyne, United Kingdom, 2010.

[PAP 12] PAPARODITIS N., PAPELARD J.-P., CANNELLE B., *et al.*, "Stereopolis II: a multi-purpose and multi-sensor 3D mobile mapping system for street visualisation and 3D metrology", *Revue Française de Photogrammétrie et de Télédétection*, vol. 200, pp. 69–79, 2012.

[PFE 07] PFEIFER N., DORNINGER A., FAN H., "Investigating terrestrial laser scanning intensity data: quality and functional relations", *Proceedings of the 8th Conference on Optical 3-D Measurement Techniques*, Zurich, Switzerland, pp. 328–337, 2007.

[POM 13] POMERLEAU F., COLAS F., SIEGWART R., *et al.*, "Comparing ICP variants on real-world data sets", *Autonomous Robots*, vol. 34, no. 3, pp. 133–148, April 2013.

[REM 11] REMONDINO F., "Heritage recording and 3D modeling with photogrammetry and 3D scanning", *Remote Sensing*, vol. 3, no. 6, pp. 1104–1138, 2011.

[RES 09] RESHETYUK, Y., Self-calibration and direct georeferencing in terrestrial laser scanning, PhD Thesis, Department of Transport and Economics, Division of Geodesy, Royal Institute of Technology (KTH), Stockholm, Sweden, 2009.

[RUS 01] RUSINKIEWICZ S., LEVOY M., "Efficient variants of the ICP Algorithm", *Proceedings of the 3rd International Conference on 3D Digital Imaging and Modeling, 3DIM*, pp. 145–152, 2001.

[SCH 07] SCHULZ T., Calibration of a terrestrial laser scanner for engineering geodesy, PhD Thesis, ETH Zurich, Switzerland, 2007.

[SMU 62] SMULLIN L., FIOCCO G., "Project Luna See", *Institute of Electrical and Electronics Engineers Proceedings*, vol. 50, pp. 1703–1704, 1962.

[SOU 11] SOUDARISSANANE S., LINDENBERGH R., MENENTI M. *et al.*, "Scanning geometry: influencing factor on the quality of terrestrial laser scanning points", *ISPRS Journal of Photogrammetry and Remote Sensing*, vol. 66, no. 4, pp. 389–399, 2011.

[VÖE 08] VÖEGTLE T., SCHWAB I., LANDES T., "Influences of different materials on the measurements of a terrestrial laser scanner (TLS)", *International Archives of the Photogrammetry, Remote Sensing and Spatial Information Sciences*, vol. 37, Beijing, part B5, pp. 1061–1066, 2008.

[WAN 08] WANG Z., BRENNER C., "Point based registration of terrestrial laser data using intensity and geometry features", *The International Archives of the Photogrammetry, Remote Sensing and Spatial Information Sciences*, Beijing, vol. 37, part B5, pp. 583–589, 2008.

[WEH 99] WEHR A., WIEDEMANN A., "Fusion of photogrammetric and laser data", *Proceedings of the 17th CIPA Symposium*, Olinda, Brazil, 1999.

[WIL 03] WILLIAMS N., HANTAK C., LOW K., *et al.*, "Monticello through the window", *Proceedings of the 4th International Symposium on Virtual Reality, Archaeology and Intelligent Cultural Heritage (VAST 2003)*, Brighton, UK, 2003.

4

Complementarity of Techniques

4.1. Introduction

Photogrammetry and lasergrammetry have been discussed one after the other in the previous chapters of the text. It can be useful to compare them (see also [GRU 08, AND 13, RAS 13]) and to show an example of how these two techniques benefit from being used together. Only the phases of data acquisition and production will be addressed here, the cloud point processing phase is discussed in detail in Chapter 5, and remains mainly unchanged regardless of the source of the data.

4.2. Comparison of techniques

The two techniques described above can be compared from different viewpoints:

4.2.1. *Data acquisition in the field*

– *Equipment:* the average equipment used for laser digitization is more expensive than that used for photogrammetric surveys. The logistics that are required to use a laser scanner are also more burdensome (equipment can be heavy and needs an electricity source). Moreover, the data acquisition process with laser scanners is quite long (compared to the almost instant photogrammetric process), and it is therefore essential to work on perfectly stable ground. The cameras which provide the photogrammetric information of the site can be easily used with a number of different devices (masts, poles, ground-based and airborne vehicles) in order to tackle any possible site layout.

– *Lighting and rendering:* lighting is a delicate matter for photographic images, whereas laser digitization could theoretically be carried out without any light.

– *Materials:* lasergrammetry is effective on the majority of materials, whereas photogrammetry (and particularly automatic methods) is ineffective on non-textured, shiny and reflective surfaces and works poorly on structures, which allow light to pass through them, such as the Eiffel tower.

4.2.2. *Quality of data*

– *Color:* images naturally provide a color rendering, which is not always present with laser scanners or at least is of a lower quality.

– *Faults:* data obtained by lasergrammetry have no or few faults, in the sense that everything which is in the equipment's field of view is recorded. Photogrammetric methods of dense image matching which produce data of the same type as lasergrammetry can generate false reliefs (incorrect matching due to defective material or repetitive patterns).

– *Accuracy:* the accuracy of laser scanners is relatively constant throughout the equipment's range. In photogrammetry, it depends on the images' spatial resolution and the stereoscopic configuration of the survey (combined effect of B/H ratio and the number of images where the points are visible).

4.2.3. *Data processing*

– *Georeferencing:* the registration of a laser station can be almost immediate if tie points and ground control points have been captured during the field-based acquisition. For images, processing is slightly longer (mainly due to the step of automatically generating tie points).

– *Production of 3D models:* whatever methods are used to obtain the 3D models (manual stereoplotting or processing of point clouds from lasergrammetry or photogrammetry), the production times are all very similar.

4.3. Example of the survey of Amiens Cathedral

Amiens Cathedral, a gothic masterpiece built in the 13th Century, is a particularly outstanding building, due to its size, fractal decoration and symbolic importance. Many of its structures are well-known due to their extremely fine detail, particularly on the entrance to the South transept and on the Western façade. [CHA 06] describes it as being "like a giant screen of lace"! He also talks of a "sort of shrub teeming with gables and pinnacles". It is a UNESCO world heritage site, and therefore deserves a complete geometric survey, which is faithful to its shapes and decorations. The Modeling, Information & Systems Laboratory at the University Picardie Jules Verne started the e-Cathédr@le[1] project whose objectives are to digitize the structure in its entirety and to provide tools to manipulate the model [HÉN 11].

1 https://mis.u-picardie.fr/E-Cathedrale/

4.3.1. *Proposed surveying methodology: lasergrammetry for the framework, complemented by photogrammetry*

The framework of the 3D model (interior and exterior) is obtained by lasergrammetry due to terrestrial laser scanners, which provide relatively dense point clouds (from 5 mm for notable details to 2 cm for relatively flat walls). The areas which the laser scanners struggle to digitize (e.g. lower part of rose window hidden at ground level by a balustrade, roofs) are processed by photogrammetry, as well as the areas which are particularly heavy on detail or colors which require high-resolution photographic images (sculptures on doorways and statues).

Therefore, to complete the hidden parts of the rose window, the accessible ledges can be used to take very close-range photographs of the façade, by fitting a camera with a very short focal length lens to a pole. This image surveying using the pole will be processed by dense matching to complete the point clouds acquired by lasergrammetry (Figure 4.1).

Figure 4.1. *Survey of the lower part of the rose window (left) and shaded image of the corresponding relief (right)*

As for the roofs, they are photographed at very high resolution from a glider or a small aircraft authorized to fly at low altitude. Highly redundant stereoscopic images are processed by dense matching and provide a 3D point cloud (Figure 4.2). However, it should be noted that the complex shapes of the roofs make it difficult to achieve the same resolution as obtained during the ground-based image acquisition. This can complicate the process of merging the different pieces of data.

Aerial Images — *Shaded relief image of the Cathedral and its surroundings*

Figure 4.2. *Photogrammetry of the upper parts of Amiens Cathedral*

4.3.2. *Georeferencing of data*

In order to successfully work with these different types of acquisitions, it is necessary first of all to coherently merge the data together. Various means of positioning are used to georeference all the data within the same system: the use of GNSS to link to the national reference system, densification via traditional topographic methods and image or laser matching over shared areas. This enables, for instance, the point clouds calculated with aerial images to be naturally placed adjacent to clouds created from the ground by lasergrammetry. For a building like this, a local reference coordinate system which is adapted to the structure's architecture may also be a solution (with a formula to link the various coordinate systems).

Despite the collection of different data within the same system (3D point clouds obtained by images and lasergrammetry), we should not under-estimate the difficulty of merging caused by the inevitable deviations in georeferencing and varying resolutions. These issues will be addressed further in Chapter 5.

4.3.3. *Completeness of the survey*

The complexity of the building makes a complete laser digitization impossible, both from the exterior and interior. The appearance of the fractal decoration, displaying great levels of new detail, is very hard to record even for the most optimistic. However, if specific requirements for comprehensiveness over an area are given, complementary work with photogrammetry is the simplest solution, given that the equipment is much less cumbersome. One option that may be taken is to carry out a survey which links with existing georeferenced data (from images or laser). The georeferencing of these images can be calculated by resection on existing data (creating new ground control points common to both sets of data). It is then possible to process the new images.

4.4. Bibliography

[AND 13] ANDREWS D.P., BEDFORD J., BRYAN P.G., "A comparison of laser scanning and structure from motion as applied to the great barn at Harmondsworth, UK", *Proceedings of the* 24th CIPA *Symposium*, Strasbourg, France, 2013.

[CHA 06] CHASTEL A., *L'Art français – Pré-Moyen âge – Moyen âge*, Flammarion, Paris, 2006.

[GRU 08] GRUSSENMEYER P., LANDES T., VOEGTLE T., *et al.*, "Comparison methods of terrestrial laser scanning, photogrammetry and tacheometry data for recording of cultural heritage buildings", *The International Archives of the Photogrammetry, Remote Sensing and Spatial Information Sciences*, vol. 37, part B5, Beijing, China, 2008.

[HÉN 11] HÉNO R., CHANDELIER L., CLAUZEL L., *et al.*, "Complementarity of terrestrial laser scanning and DSM from aerial photography: the example of the cathedral of Amiens", *Proceedings of the 23rd CIPA symposium*, Prague, 2011.

[RAS 13] RASZTOVITS S., DORNINGER P., "Comparison of 3D reconstruction services and terrestrial laser scanning for cultural heritage documentation", *Proceedings of the 24th CIPA Symposium*, Strasbourg, France, 2013.

5

Point Cloud Processing

Surveying methods using photogrammetry or lasergrammetry can therefore produce digital representations of sites of interest in the form of dense 3D point clouds, with a given accuracy and completeness. Although this type of model can then offer spectacular visualizations of the structure almost instantaneously, these point clouds are often difficult to use as they are from an operational point of view. They are objects which contain a large volume of information that has not been interpreted, and from which it is sometimes difficult to extract easily readable data. In practice, it is rare for the digital building specifications to end at this stage, unless subsequent processing steps have already been planned in the company commissioning the project, or if a different service provider is contracted, or if the objective of the project is to create a digital archive of the building. Further operations are usually necessary to be able to effectively use the abundance of information, which enabled the data acquisition.

5.1. Visualization and organization of 3D point clouds

5.1.1. *Ways of visualizing point clouds*

The visualization of data acquired is a key element for the effective and reliable use of 3D point clouds. Part of the quality evaluation of this data is often carried out by direct visualization in specially designed software. This allows the user to obtain a large amount of information, especially on the completeness of the survey and the errors or noise in the measurements. Although a quantitative evaluation of the digitization and processing carried out remains essential because it ensures objectivity, these qualitative aspects can often help to quickly identify and correct any defects as soon as possible. In practice, the rendering method used should make the data as easy to read as possible and it is recommended to test a number of visualization methods to identify those which are the most relevant according to the type of work carried out.

5.1.1.1. *Visualization in the form of a depth image*

The acquisition geometry of the two surveying methods – image and laser – provides a natural visualization of the 3D point cloud in the form of a depth image (2D raster image for which the value associated to each pixel represents a distance):

– in the case of dense image matching, the 3D point search is carried out on a 2D grid, attached either to the master image or to a plane linked to the object;

– the spherical acquisition geometry of the laser scanner allows for an almost immediate 2D representation via a spherical projection. The depth can be set by the intensity of the returned laser beam or the distance ρ at the center of the instrument's spherical coordinate system.

a)

b)

Figure 5.1. *Examples of visualizations of 3D point clouds in the form of a raster image of a) orthocylindrical projection and b) spherical projection*

This type of 2D visualization is well adapted for matching results for a single master image, a single-laser station or when the object or part of the object in question can be interpreted in 2D either directly (e.g. for a fully plane façade) or via a simple geometric transformation (e.g. for the spherical projection of a dome). The depth map is therefore an efficient and intuitive visualization method, which is compatible with the screens commonly used by processing software.

If we use a different 2D visualization plane to that which is naturally created by acquisition geometry, it is then necessary to specify the regular grid used. This is determined by the position and orientation of the grid plane in the space as well as by the position of the grid's origin in the plane and the grid step size. Furthermore, the nature of the depth to attribute to the grid nodes, distance or parallax, must be established as well as the calculation method for the depth values of each node. A number of interpolation methods can be used: those traditionally found in image processing (nearest neighbor, average or median value, etc.) or by using a mesh constructed from the point clouds.

Once the values of the 2D image are known, several types of display can be used to represent them: grayscale, hypsometric tinting or shading (Figure 5.2). Each method lends a specific lighting to the data and can allow the emergence of elements of interest [BRY 13]. For instance, the shading method emphasizes the noise levels in the data. A combination of the various methods can also be carried out.

a)

b)

Figure 5.2. *Depth map visualization methods: a) with shading (Amiens Cathedral – Arcade of kings) and b) mixing shading and hypsometric tinting (Forteresse de Salses, [PIE 13])*

5.1.1.2. *3D Rendering*

Point cloud operating software uses 3D rendering libraries; the two main examples are OpenGL[1] and DirectX[2], in order to ensure the 3D visualization. The user is able to navigate through the data, zoom in and out, and rotate around the objects.

Different display modes are usually available. By combining these possibilities, the user can find the best way to consult and understand the data:

– coloring of points depending on values of intensity, RGB color or distance, according to the source of the data (which laser station or master image) and, possibly, by applying processing to improve the appearance and make the data easier to read (stretching, gamma correction or any other operation designed to improve the image);

– sub-sampling of data during navigation in order to lighten the manipulations necessary;

– modification of point size;

– orthogonal or central projection;

– use of specific renderings designed to make individual parts of the object stand out.

On this final point, there are a number of methods in the field of 3D computer graphics, some of which are available in point cloud processing software. However, they are more generally used to represent meshing (see section 5.3.3) from these clouds. These 3D renderings can be photo-realistic by attempting to produce a depiction, which is as faithful as possible, and consist of [MAL 05]:

---

1 http://www.opengl.org.

2 http://msdn.microsoft.com/fr-fr/library/dd144146.aspx.

– a lighting method which defines the illumination of objects according to the light sources, which are present and according to the interaction between the objects either by local (Gouraud, Phong algorithms) or global calculations (radiosity method);

– a visibility calculation method: the most common is the Z-buffer method, which enables the processing of depth and hidden parts of the scene by creating a map which can be superposed on the 2D view and where each pixel contains the distance between the observer and the closest object in the scene.

The traditional ray tracing algorithm has these two parts. It consists of simulating the opposite path of the light, starting from the observer: the properties of reflection and diffusion of objects are taken into account, as well as the obstacles encountered by the beam, which can then help to determine the hidden parts of the scene.

Some rendering methods try to reproduce the lighting conditions, as closely as possible, from the time when the building was constructed [HAP 09].

Non-photo-realistic rendering methods often have an artistic objective or, more specifically in our case, contribute to making the object data easier to read. Those methods, which accentuate the contours and silhouette of the structure, are particularly useful (Figure 5.3). To do this, they use the information of the surface normal of each point. This can be calculated by estimating the tangential plane of the cloud on the neighboring point with least square adjustment or also by performing a principal component analysis on the neighboring points. [DEY 05] and [KLA 09] suggest comparisons between existing methods. The variations can have an effect on different aspects: optimization of calculation time, insensitivity to cloud noise etc. In order to make navigation more fluid and since it is used for several aspects of processing, the surface normal can be calculated once and for all and recorded at the level of each point.

In practice, not all the methods provide a real-time calculation and we might consider using static 2D images for some parts of the analysis if the most appropriate rendering method does not allow for a dynamic visualization (see also section 5.4).

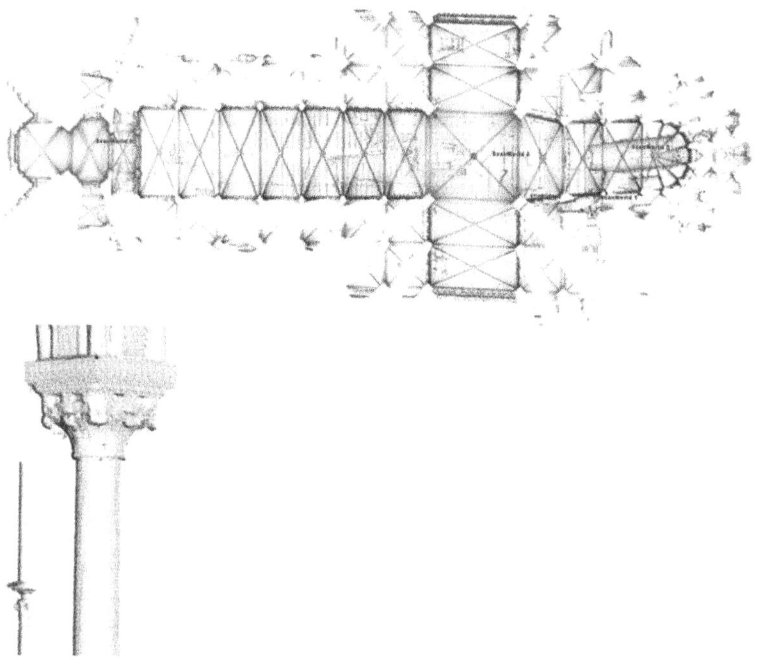

Figure 5.3. *Visualization in silhouette mode – Saint-Denis Cathedral*

Finally, if the original point cloud is displayed as it is in the software, edge-aliasing effects may emerge when it is visualized from close- or long-range. Adapted interpolation methods exist to limit these effects (on this subject, also see section 5.2.3).

5.1.2. *Organization of point clouds*

5.1.2.1. *Strategies for processing large volumes of 3D data*

For large structures, it is rare to need access to all of the data, to visualize it and to process it simultaneously. To avoid having to process more data than necessary, a practical and relatively natural process is to define the subprojects or layers, which clearly correspond to separate parts of the structure (level by level, from the inside and outside, room by room, etc.). This process can be a little time-consuming because this type of operation requires manual selection, but often time is saved during later processing stages, since the effectiveness of calculations and navigation has been improved. It is advisable to define these basic subsections with the project sponsor who usually has perfect knowledge of the structure and will therefore know of any existing grouping and naming systems that may be suitable for the organization of digitized data. Organization methods may have been defined on a local or national level and implemented on other sites [PÈR 10], and it can therefore be useful to use these as a guide.

For software specially designed for the processing of 3D data, divisions into different parts, regardless of the content and significance of the information, are often used to improve efficiency. The access to data, which helps to increase the amount of processing and to simply manage different detail levels, is hence optimized. Several types of divisions are traditionally used in the fields of spatial indexing, nearest neighbor searching, detecting collisions and in a more general way, in 3D computer graphics:

– partitioning using a regular grid cut up into groups of identically sized voxels (cubes, 3D equivalent of pixels) like in a 2D image: the simplicity of this type of structure and the speed of the data flow are countered by the many empty cells, which remain and the problem of choosing a scale for the grid;

– partitioning using a hierarchical or adaptive grid, which cuts up the scene so that no cell contains more than *n* elements or until the maximum number of predefined levels is reached. This type of organization is a good way of adapting to the irregular distribution of data and makes the storage and use of large volumes of data easier [ELS 13]. The main methods involving hierarchical grids are:

- octrees, which are generalizations of 2D quadtrees in 3D, (Figure 5.4): using a voxel, which contains all the data, the area is divided up repeatedly and hierarchically into eight subcubes until the chosen end criteria have been reached;

- binary space partition (BSP), whose only principle is that space should be divided up into planes of any kind. Kdtrees are a particular case whereby planes are chosen parallel to the axes of the coordinate system [GOS 13];

– partitioning using the bounding volume hierarchy method, well-suited to moving objects.

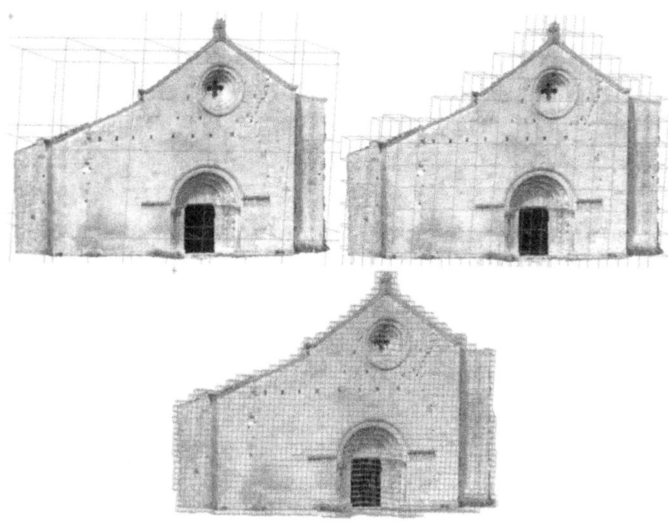

Figure 5.4. *Octree structuring: point cloud and different levels of the hierarchical grid*

Whichever division method is used, the list of data (points or polygons) belonging to each tree node (cell of a certain level) is stored in order to optimize the access to data. The most commonly implemented technique by software is the octrees method. Many processes described in this chapter use this type of space-partitioning data structure.

5.1.2.2. *Neighborhood information: advantages and method*

It is often necessary to know the nearest neighbor to each point in order to estimate the surface normal, initiate a segmentation process, subsample data, etc.

As previously discussed, point clouds are sometimes preserved in their acquisition geometry, which directly provides topological information. The neighbors in an image are the 4 or 8 pixels, which surround the element in question, known as 4 or 8 connexity. Since a laser scanner works vertically, the neighbors of a point are those which are just before and after it in a vertical direction and on adjacent lines, neighbors are those points on the same increment.

However, in many cases, the site needs to be addressed as a whole or at least a subsection, which corresponds to several acquisition geometries. In this case, data is transformed according to a list of unstructured points (*X*, *Y*, *Z*), where the concept of topological relationships is lost. There are techniques for recreating connections between elements, especially by using the notion of neighborhoods [DEM 11]. The division described above can be used given that the points belonging to the same cell are neighbors. For a mesh of point clouds, it is also relatively simple to calculate neighbors if the points linked by one or more parts of the meshing framework are considered as neighbors.

5.1.2.3. *Selection of 3D data*

Selection tools are necessary for manual operations on the data to be carried out. In practice, in point cloud processing software, traditional CAD functions are often present: point by point selection, by rectangle and by polygon, etc. Although some software enables 3D stereoscopic visualization, to our knowledge, this is not associated with the "true" 3D selection tools. Often, the selection stage is therefore carried out in 2D: all points which are inside the drawn area, according to the projection used for the display, are retained. It is then necessary to go through this process a number of times in order to obtain the desired selection and the operator must check the validity of the selection from a range of different angles. Indeed, hidden parts could mask the presence of points, which are not wanted. Some software enables the user to define the bounding boxes whose limits are moved and redrawn gradually by turning around the object.

5.2. Preprocessing of the point clouds

5.2.1. *Denoising and filtering*

In sections 2.5.5 and 3.5, we discussed how point clouds can suffer from the following defects:

– point cloud noise varies greatly depending on the source of the data: from less than one millimeter to several millimeters (Figure 5.5);

– artifacts or false points (outliers);

– elements to be filtered out, because they are not part of the object of interest: pedestrians, cars and dust, etc.

There are few automatic processing facilities for the last defect and it is often necessary to manually process the data in order to eliminate these errors with a selection process. This phase is made longer if the acquisition failed to properly take into account the areas specified by the project sponsor. Once again, if possible while on site, the operators in charge of the survey would be advised to consider these later processing steps by only digitizing what is contained in the building specifications and by organizing the data beforehand. This would then reduce the amount of work in the processing phase.

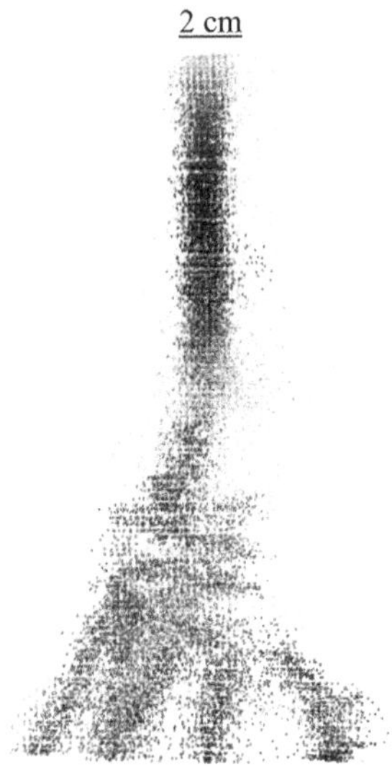

Figure 5.5. *Example of significant noise on the profile view of a target: the standard deviation for a point cloud at target level is 8 mm*

In order to reduce the noise present in point clouds, there are data smoothing algorithms that can be used. These methods can be divided into two main groups:

– methods used with traditional algorithms for denoising 2D images and adapting them to depth images: for instance, [SMI 11] applies wavelet methods and non-local means to small objects; in [SMI 13], the technique of compressed sensing is discussed, which consists of simplifying the cloud as soon as the data is recorded;

– methods working with the 3D point cloud: this can involve filters, as in [DIG 12] who carries out denoising on descriptors calculated at each point or on least square estimates for the smoothed surface, which best corresponds to the point cloud. The moving least square method is introduced by [LEV 98] and continued and expanded upon by [DES 10].

False points and artifacts can be filtered out by studying the distribution of their neighbors in space and by establishing a threshold of the average distance from the point to its neighbors as in the Point Cloud Library[3].

Depending on the situation, these processes can be applied as soon as the acquisition is carried out, or later, during the various processing stages. Evidently, the difficulty is in reducing the noise without losing the information, which is present in the data set. There are algorithms, which work directly on meshed data.

5.2.2. *Merging of data*

3D digitization carried out by laser scanner or from images leads to the collection of data, which, when assembled, provides a continuous and complete theoretical representation of the object of interest. Above, we discussed how the positions and orientations calculated during the registration phases have a certain level of accuracy, which depends on the means used to carry out and verify the georeferencing. Moreover, to ensure the continuity of the representation, overlaps are taken during the acquisition between the various measurement stations. The consequence of this is that the global representation of the object can suffer from a number of defects:

– irregular density between the point clouds because of the overlapping areas between various data sets: this type of defect can be corrected using uniform resampling methods on the point clouds;

3 http://www.pointclouds.org.

– gaps between data from different sources: tie points can be used between different entities or optimization of the distance between the point clouds, for instance by using the ICP method, which can help to directly reduce these effects during the orientation phase.

Problems encountered can be more significant if different acquisition technologies (laser vs. image) have been used and also if different types of equipment have been used within the same technique. Nevertheless, the parameters at the acquisition phase cannot compensate for all gaps, thus the question of merging is always relevant.

Depending on the desired use of the digitization and the project's building specifications, it may be necessary to correct these effects. [CUR 96] recalls that there are many expectations regarding data merging algorithms: relative inaccuracies in the input data to be addressed (use of correlation coefficient, angle of incidence, variation of depth between neighbors, etc.), use of redundant information to improve the data set (e.g. by reducing noise), independence of data integration order, ability to add new data, compatibility of calculation time with the computer facilities available, insensitivity to outliers, ability to reproduce any type of object, ability to fill in those areas without information.

In the past, 3D data merging has mainly been processed for individual objects (statues, furniture) or coherent subgroups (doorways, pillars and towers). A distinction between the algorithms can be made regarding whether the topology of 3D point clouds has been taken into account or not. Moreover, there are two calculation strategies: volumetric methods, which work in 3D (Figure 5.6), and surface methods for 2D.

a) b)

Figure 5.6. *Point cloud merging: a) registered point clouds and b) merged with texture [GAR 09]*

For larger collections of data, such as point clouds for a building's sizeable structures, this problem can be overcome by dealing with the gaps between the clouds at the registration step and by processing over-density in overlapping areas by regular spatial sampling (see section 5.2.3). This final processing step may potentially be obligatory because of the variations in resolution due to acquisition methods used.

5.2.3. *Subsampling, resampling and compression of point clouds*

There are many reasons why we may feel it necessary to rework the raw 3D point clouds: irregular density of clouds due to the characteristics of the acquisition geometry, inadequate density for the shape of the object (an almost perfect, closed plane can be defined by only its four corners) and finally, the need to have reduced data in order to carry out certain operations (to visualize the whole of the structure, carry out intermediary calculations which do not require a display of all the data). In order to satisfy these requirements, different types of operations can be distinguished according to the following terminology:

– subsampling consists of extracting a subsection of points from the initial cloud; the points, which are retained, are not modified;

– resampling consists of creating a cloud of new 3D points calculated using the initial points;

– compression involves the search for a computer algorithm, which reduces the volume of the data, with or without a loss of information.

In the case of outstanding sites, where the volume of data is very high, these phases can become essential to enable smooth data processing.

If maintaining the elements that contribute to the feature lines of the object is not a priority, a very simple subsampling will consist of retaining one point every n points. For data from a laser scanner, acquisition is generally carried out in a vertical direction. One method is to keep one point every n points in a line and one line in n. For a depth image, traditional subsampling algorithms for image processing can be reused. Another option is to subsample the point cloud by keeping a number of points as predefined by the user randomly. The limits on these simplistic methods, such as the disregard of the variations in the complexity of the objects or the need to work on a cloud of ordered points, mean that they are only used, in practice, in cases where accelerated visualization is required.

To take the spatial organization of points in the cloud into account, an octree structure can be used (see section 5.1.2). This provides various levels of data subsampling according to the depth of the tree. The subsampled cloud is then made up of the point, which is closest to the center for each leaf of the tree.

Resampling is possible by keeping the center of the leaf or the points' center of gravity present inside the cell (Figure 5.7).

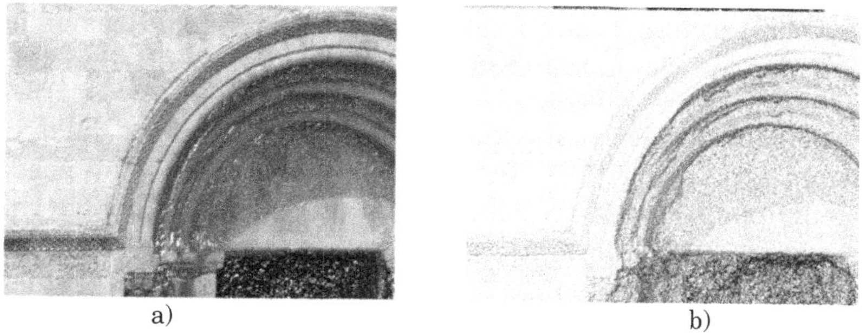

Figure 5.7. *Resampling of a point cloud: a) initial cloud and b) cloud resampled by the octree method at level 10*

Other sampling methods conserve the distance between points and therefore allow a uniform distribution of points in a space [PUT 13].

Further developments still need to be made in order to adapt the density according to the complexity of the object. This is a function that, to the best of our knowledge, is not yet offered by currently available software.

The computer programming strategy of data may complement the data compression in order to obtain a volume of data, which is compatible with the storage and diffusion constraints that may be present. The most common formats are shown in section 6.2.3.

5.2.4. *Segmentation*

The segmentation of a 3D point cloud consists of creating subgroups of coherent points. The criteria for such coherence may be geometric (grouping using simple shapes: planes) or semantic (division into structurally independent entities, for instance: rooms, statues, furniture). The segmentation phase usually comes before the reconstruction phase: by isolating the objects to be processed from exterior elements, which can cause interference, automatic processing is made easier as well as point cloud manipulation and drawing.

Segmentation is a widely studied topic in the field of image processing [MAÎ 08] and the developments proposed for 3D point clouds are often adaptations, which involve the available 3D information. Therefore, it is possible to distinguish between edge-based methods (for which segmentation is carried out inside shapes contained within an automatically detected edge), and region-based methods (which look to define homogeneous regions), or combinations of the two.

The region-based methods are more common and better adapted to 3D point clouds. As far as bottom-up approaches are concerned, these are known as region-growing methods: from an original surface calculated using a seed-point, the points belonging to the pre-defined surface are chosen according to a given criteria. This criteria generally consists of a number of elements amongst the following: coplanarity, variation of the surface normal, proximity, color and intensity. The differences between the algorithms suggested (e.g. [VOS 04] and [RAB 06]) depends on the way the seed surfaces are calculated and on the growth criteria, which constitute sensitive points and have a significant effect on the quality of the results.

The top-down method of split and merge has also been adapted to point clouds: [WAN 04] uses an iterative partition according to the octree, where the division ends when the 3D points within a leaf of the tree correspond to a given criteria, giving way to a phase of merging together the sets of data obtained in this process.

These iterative methods have the advantage of being relatively fast and able to process large volumes of data.

Some algorithms rely on an analysis of the points' features, with a view to partitioning the data via clustering methods. With this in mind, [BAR 13] suggests a method that carries out segmentation by mean shift [CHE 95] by combining the parameters of radiometry, depth and the surface normal via a majority vote system.

Another large family of segmentation algorithms is made up of "direct" methods, which usually aim to extract planes from data. These are often 3D generalizations of the Hough transform with the use of an adapted parametric space [BOR 11] or RANSAC-type algorithms [FIS 81], which are better at taking outliers into account [BOU 07].

The state of the art in the field of ground-based acquisition mainly focuses on the processing of surveys from mobile laser scanners: indeed, the objective of these systems is the reconstruction of street views by complementing information obtained by aerial images. There is a clear requirement for automation in order to provide geometric models on very large scales within reasonable production times. In this context, the entities to isolate are generally quite simple, and are mainly planes: façades (Figure 5.8), roofs and possible elements, which make up buildings such as doors, windows and balconies. Sometimes, elements that can disturb the reconstruction process (pedestrians, vehicles and street furniture) are also searched for [HER 09].

In practice, there are fewer examples of segmentation algorithms adapted to the more complex buildings, and able to isolate more detailed objects that simple geometric shapes previously stated. [DEV 05] suggests a method with the combined use of laser data and the external image of a church: a hierarchical segmentation algorithm is applied to the image then geometric information is introduced into the energy function, which describes the way the segments of the first stage are merged. [ZHA 09] applies the region-growing method with criteria combining neighbors and color in the case of a traditional Chinese house. The common aspect shared by these methods is the combination of geometric and radiometric information in order to optimize the response to the problem faced [AIT 13]. Prior knowledge of the object to be isolated can also be beneficial: existence of a privileged direction (vertical in the case of a building), symmetry and repetition.

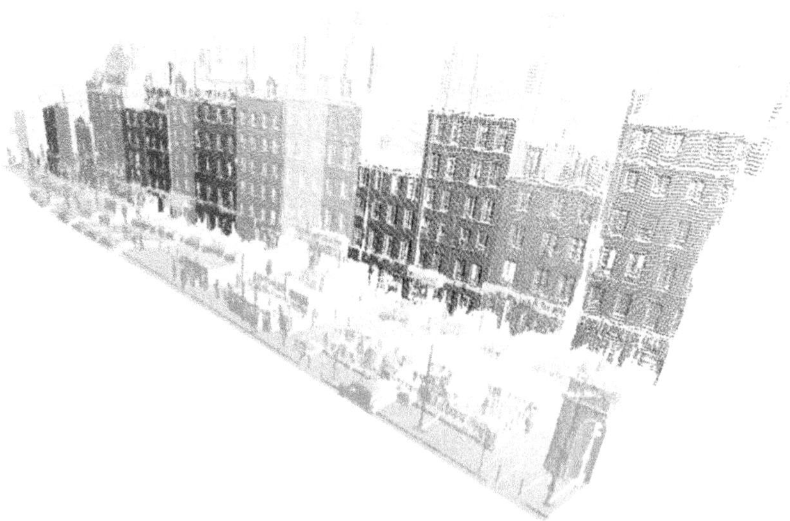

Figure 5.8. *Segmentation of a point cloud from a mobile laser scanner [HAM 11]*

The results of the automatic segmentations are evaluated by quantifying the number of over-segmentations (too many segments) or under-segmentation (grouping of objects of different types). Finally, a complementary step of semantic attribution may be necessary in order to give meaning to the extracted information.

5.3. From the point cloud to the 3D geometric model

The goal of this section is to explain how it is possible to go from a model in the form of a 3D point cloud to a 3D geometric model, which enables the continuous representation of the site of interest in space. For the sake of completeness, the model should also contain semantic information on their nature and their relationships to one another. These areas will be discussed in Chapter 6.

5.3.1. *Types of 3D models*

There are different types of 3D models, which are more or less adapted to the representation of outstanding sites:

– solid model: assembly of elementary volumes (cubes, spheres, cylinders, etc.) according to set operations (union, intersection, difference, etc.);

– boundary representation, as wire-frame models or polygonal meshes: a representation of the object by collections of lines or surfaces. This can be a simple line or more complex geometric objects such as a NURBS, a B-spline generalization;

– other types of models, which are less commonly used, as a final product: sets of surfels, voxels, etc.

In practice, for complex structures, if the aim is to represent the object "as it is" with a minimum amount of simplification, it is often boundary-based modeling, which is used. This may be a mesh composed of cells (often triangles) connected to each other by their vertices (often points belonging to the cloud). There are many automatic triangulation algorithms, which directly enable the creation of this type of 3D model from point clouds (see section 5.3.3). The advantage here is that it is possible to produce a relatively faithful representation of the original object while at the same time staying close to the points surveyed. The disadvantage lies in the fact that the model that is obtained in this way can be quite heavy since the data to be conserved will be multiplied. In this case, having input data of a point cloud whose density is adapted to the object is often essential. Moreover, the errors committed by the automatic algorithms sometimes need correcting manually, which can be a rather time-consuming job.

Another way of using segments for modeling, as shown in section 2.6, is to represent the object with polylines, which correspond to the characteristic elements of the object: edges, frameworks, material limits, etc. Some automatic extraction algorithms exist for simple buildings (see section 5.3.4) but they are often insufficient in the case of complex structures. As a result, it is generally necessary to earmark a manual drawing phase for the point cloud (see section 5.3.2).

Furthermore, work is being carried out to create a 3D architectural model adapted to the case of more complex structures and which can be designed with an analysis of the construction rules and models linked to the type of architecture in question [DEK 03, DEL 07, CHE 09].

Finally, modeling using volumes should not be definitively forgotten about as they can provide a suitable solution if, for example, the goal of the project is to make the model available to the general public.

5.3.2. *Modeling in a 3D point cloud*

As stated in the previous section and despite research carried out in recent years (see section 5.3.4), 3D modeling of a complex structure is still done mainly by manual plotting. This phase is extremely time-consuming: for instance, [STE 04] estimates that three quarters of the overall project time, from acquisition to delivery, is dedicated to the plotting phase for a site of architectural interest. This phase requires capacities of abstraction, for example, when it is necessary to draw irregular objects (statues, cornices) or curved objects (vaults, columns) with linear tools. The use of regular geometric shapes, possibly predefined, will tend to provide approximations, which are not always appropriate due to the weakness of the catalogues generally made available.

5.3.2.1. *What to plot and how?*

This question is simply reiterated here, because it fundamentally affects the plotting work that needs to be carried out. In any case, it will have already been addressed during the creation of the building specifications and the consideration of the acquisition parameters so that, at this stage, it is possible to identify and plot the required elements.

Once again, depending on the building specifications, either 2D or 3D products may be requested, which has a direct effect on the type of methods to be implemented. As discussed earlier, the rendering and segmentation methods make the drawing process easier and avoids errors by giving the operator the data necessary with a suitable visualization. Capture operations are often carried out by CAD software, with a specific plug-in for coping with large volumes of data to make navigation within the point cloud feasible and ergonomic. The geometric shapes used for these operations can often be traditional shapes (polylines, arcs, splines) or, as previously discussed, they may also be specifically designed parametric models.

Another legitimate question regards how to complete the plotting operation when there is information missing. Although [CHE 09] points out that this type of complementary work can be carried out by experts, it is important to understand that the objectivity offered by the techniques described here cannot be respected in this case. It is therefore advisable to save the origin of data in an attribute.

5.3.2.2. *2D plotting*

Occasionally, requests are made for 2D representations of a site (see section 5.4), which may seem like a shame given the mine of information available in 3D. This clearly shows the current lack of manipulation and interpretation tools for 3D data. Furthermore, as discussed in the following paragraph, the means to be implemented in order to obtain truly 3D data are significantly heavier and more expensive.

The 2D plotting starts by defining the work plane and then consists of simply tracing on the plane the elements of interest using a slice of the point cloud (Figure 5.9). Snapping functions may be used to ensure that the lines drawn are as close as possible to the input data.

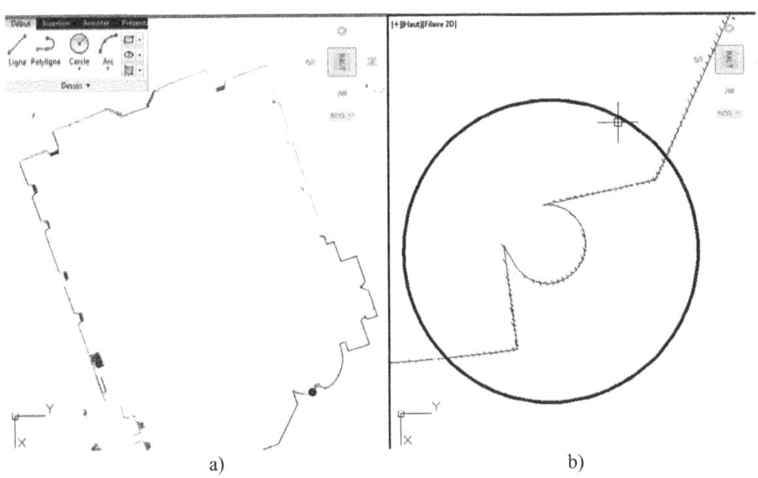

Figure 5.9. *2D plotting on a point cloud: a) view of an entire cross-section of a plane and b) zoomed-in drawing*

5.3.2.3. *3D plotting*

The first possibility for 3D modeling consists of associating several 2D drawings such as those described in the previous paragraph by varying the work plane. In order to obtain a continuous surface, it is necessary to connect the linear elements with the appropriate tools: for instance, extrusion, splitting. The advantage of this method is that it is quite simple to implement, but it does nonetheless involve relatively large approximations in the modeling due to the potential gaps between the points compared to the plane.

One way of working in 3D for real is to use snapping options, which are often available in the software tools. These enable a search for the closest point in the direction of depth, defined as the perpendicular to the viewing mode on the screen. One area of difficulty is that it is not always possible to visualize whether the snapped point is relevant and correct. Indeed, at the edge of an object, for example, it may be the case that the point selected is behind the object, without this being picked up. To avoid this from happening, the number of viewpoints should be increased by turning around the object, which can be a little arduous, and by pre-adjusting in the interface several viewpoints simultaneously to check that the elements drawn are correct.

Finally, there are intermediary techniques, which show that depending on the tools available and the modeling required, it may be necessary to define a specific working method, which is as ergonomic and effective as possible [BAL 13]. In particular, the 3D capture based on parametric shapes, which consists of attaching the chosen objects from the library to the 3D point cloud ([MUR 13] and the historic building information modeling (HBIM)), may well be a relevant methodology.

In any case, the computer used for this type of production must be high-performance, especially regarding the graphics card, which must be compatible with the manipulations to be carried out. This therefore generally means specially designed technical equipment. A solution that is likely to be more effective is to use a tool enabling 3D plotting with a stereoscopic visualization of the point clouds.

5.3.3. *Surface meshing*

Meshing involves a spatial discretization in a collection of different faces, often triangles (triangulation or even tetrahedrization) or polygons, such that two elementary cells are either disjointed or with a shared edge. This type of model allows the representation of an object of interest without interruption, without prior knowledge and with little interpretation of its shape and appearance. It is particularly well-adapted to point cloud processing since the vertices of the faces, if required, can be directly taken from the point clouds, which means that the connection with the original measurements is maintained. The meshing calculation and surface reconstruction using points are areas which attract high levels of interest in the field of algorithmic geometry. As a result, the references here are mainly taken from this field and the objects processed are extremely varied. Free-access tools and libraries offer this type of service[4, 5].

There are various 2D triangulations and the most well-known is undoubtedly the Delaunay triangulation [DEL 34]. This has the following property: "no point should be situated inside the circumscribing circle of any triangle". Depth images, as they are intrinsically 2D, are well-suited to Delaunay triangulation to be applied directly onto point clouds (Figure 5.10). A relatively significant disadvantage in this case is the requirement to work station by station or

4 Computational Geometry Algorithms Library (CGAL) – www.cgal.org.
5 Point Cloud Library (PCL) – http://www.pointclouds.org.

image by image and the connections are not processed. Moreover, filtering is necessary since not all the points can be connected: triangles with an unusually long side and those whose vertices connect very different depths can be eliminated simply by a threshold value.

Figure 5.10. *Meshing obtained by 2D Delaunay triangulation on laser data*

True 3D methods allow all the acquired data to be interpreted simultaneously. The strategy to adopt may differ according to the amount of knowledge available on the point cloud (surface normal, neighborhoods, organization of data according to a regular matrix).

In the field of explicit approaches – those which construct triangles directly from points – some traditional algorithms can be given, without providing an exhaustive list, such as adaptations of the 3D Delaunay triangulation [LIU 05], Power Crust [AME 01], Alpha Shape [EDE 94] and Ball Pivoting [BER 99] methods.

a) b) c)

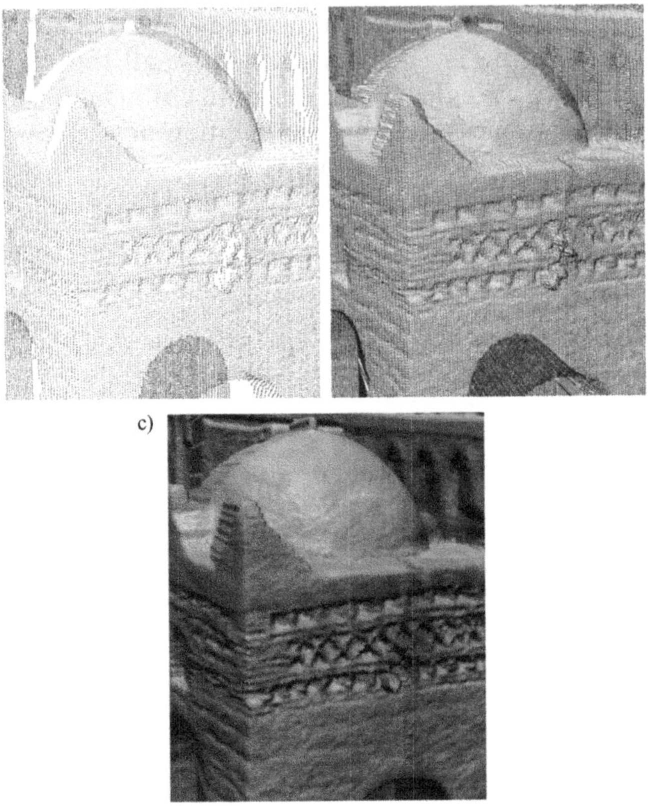

Figure 5.11. *Zoom in on Figure 5.10: a) point cloud, b) wire-frame meshing and c) shaded meshing*

If data is noised or contains outliers, despite the processing that may have been carried out, it is advisable to use implicit methods, which approximate the surface to be reproduced. These algorithms generally work in two stages: first, the surface is approximated by a function and then the meshing is generated. In this family, approaches based on Marching cubes [LOR 87], Moving least-squares [LEV 98], and Radial basis function [CAR 01] methods, or Poisson surface reconstruction [KAZ 06] can all be included.

An alternative to surface meshing proposed by [PFI 00] uses the concept of "surfels" (surface elements, the term "point splat" is also used). This is where each point is associated with a disk locally perpendicular to the cloud with a radius so that the disks overlap to form a continuous surface.

The expectations regarding this type of reconstruction concern their robustness (presence of false triangles), their ability to process extremely varied figures (different densities and objects), to manage data voids or not [FAB 07], to take into account geometric constraints (buildings' characteristic lines). In practice, it is rare to achieve a perfect triangulation, regardless of the algorithm chosen. Consequently, it is very often necessary to implement manual corrections to obtain a correct 3D geometric model.

For more complete information, refer to other works in the field of algorithmic geometry, such as [BOT 10].

5.3.4. *Automatic reconstruction*

Above and beyond automatic meshing calculations, many research projects have been carried out on the automatic reconstruction of buildings' characteristic elements using 3D point clouds, especially since the arrival of laser scanners. This means replacing the operator's manual modeling process with an automatic calculation. Of course, providing an algorithm with building specifications of the elements to survey and obtain in return a 3D geometric model is not yet a realistic possibility. Once again, the most cutting-edge results involve reconstructions of a street's façades on a city-scale because, generally speaking, the elements to survey have common characteristics that can be formalized via a common basis of knowledge which can be introduced at the start of the algorithm: verticality of walls, presence of structural elements (windows, doors, balconies), symmetry, horizontal alignment of levels, etc.

As far as outstanding sites are concerned, their architecture is more difficult to transform in to standard rules. This means that the automated product must be seen as a partial solution, which can then make the plotting easier and speed up the working process.

Automatic methods usually work by extracting sets of primitives (contours, planes, spheres, etc. [BOU 10]), which are obtained with the same processes as those discussed in the section on segmentation for shape searching (see section 5.2.4). The most common and most effective method is RANSAC, which enables the user to obtain the different types of primitives discussed above, as shown in [SCH 07]. The limit of these methods is that it is generally not possible to represent everything with primitives: gaps remain which need to be filled in.

Approaches that combine reconstruction using primitives and meshing [LAF 13] can provide solutions to the complex cases being discussed by adapting to the varied content of buildings to process (Figure 5.12).

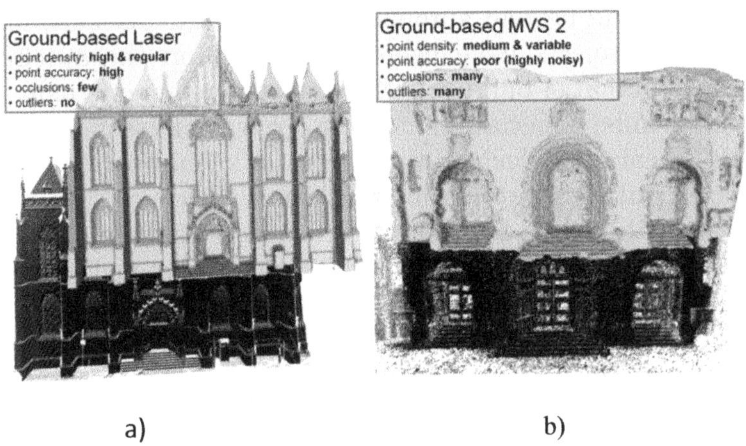

Figure 5.12. *Results of a hybrid reconstruction method, using a point cloud: a) from a laser and b) from images [LAF 13]*

It should nevertheless be noted that these algorithms require long calculation time and it is unrealistic at the current time to expect to be able to process outstanding sites simply by inputting all collected data. Breaking the data down into subareas is essential for this type of operation.

5.3.5. *Combining laser point cloud with images*

The information contained in the point clouds from laser scanners is generally excellent from a geometric quality point of view. As discussed in section 3.5.2, progress still needs to be made in terms of radiometric quality, particularly in comparison with images from photographic equipment, which is currently available on the market.

First, for photo-realistic renderings, it may be essential to supplement laser acquisitions. Second, external images also favor the operations described above, whether on the level of preprocessing or for the automatic extraction of information from point clouds.

For both types of information to be used in a complementary way during processing, it is necessary for them to be perfectly registered with each other. The usual methods used to calculate their georeferencing (see sections 2.4 and 3.4) are able to satisfy this requirement without great difficulty. However, in some specific cases, for instance when one of the data groups is used to supplement the other, it is advisable to calculate the relative registration of one of the data sets to the other. This operation can be carried out manually by locating at least three common reference points between the image and the point cloud. There are also algorithms which process this stage automatically, by extracting primitives from both types of data and then matching them [DEV 04, MEI 10, CRO 13]. Laser data can be transformed into 2D and then the traditional methods for detecting and matching points of interest developed for images can be used. Furthermore, it is wise to reproject them in a geometry, which is compatible with that of the images, so as to limit the effects of perspective. In practice, this operation is not always successful, because signals present in the two types of data may be different due to the gaps in wavelength used, different lighting at the time of the acquisition and viewpoint variations.

For the reconstruction problems discussed previously, using two types of data jointly improves the results, because the information that is used more abundant [DEV 04, BEC 07, PU 09, HAM 11]: the images often have a higher resolution, they show the objects' edges and they take color differences into account.

5.3.6. *Texturing*

To obtain a geometric model with a realistic appearance, it may be necessary to texture the 3D objects using the images. [KER 12] summarizes the different circumstances:

– if the 3D reconstruction is carried out by image matching, color information is directly available at the level of the points of the 3D model;

– texturing can be manually carried out with CAD software;

– texturing can be calculated after the 3D reconstruction, by external images whose internal geometry is known as well as its position and orientation in the coordinate system of the geometric model.

In this last case, the operation may consist of improving the color of the points in the cloud [CRO 13]. Usually, this is carried out on the mesh, whose faces need to be colored. The texture calculation for each face can be done automatically [ALS 05]. First, it is necessary to know the hidden parts between the image and the 3D model. There are two main groups of calculation methods for occlusions: from the image (e.g. Z-buffer) or from the object. Then, if a number of images are likely to overlay a given face, a rule for choosing which image to use must be defined. This can be done with criteria based on the resolution, or the angle between the image viewpoint and the surface normal. Finally, the interpolation method of the radiometry within the face in order to create a continuous colored surface can either be affine (simpler, but less rigorous) or can take into account the perspective of the image (which takes longer to calculate).

Textured 3D models (Figure 5.13) are highly sought-after because of their ease of interpretation but the stages involved in creating them can be relatively long. This corresponds to what has been discussed in the previous paragraphs and these stages become longer as the amount of detail required increases. Furthermore, it is worth noting that there is currently no single tool that can provide a full solution to this problem. In practice, specialized companies or bodies which process this data often use a combination of software: computer graphics library, 3D modelers, etc.

Figure 5.13. *Textured 3D models of Araf mosque, Yemen [GOD 09]*

5.3.7. *Qualification of geometric models*

The evaluation of the result is particularly delicate.

From a purely geometric point of view, it can be useful to map the gaps between the reconstructed models and the initial point clouds. However, the qualitative information obtained in this way does not help to evaluate the quality of the model in an absolute sense. Indeed, the point cloud is not exempt of defects: notably hidden parts and georeferencing inaccuracies. To obtain absolute data, it is advisable to carry out specific tacheometric measurements in the field, which would provide a more objective view. [LAN 12] suggests a method for evaluating segmentation and reconstruction using the definition of a set of quality indicators.

From a radiometric point of view, the evaluation is once again often made by a visual judgment, due to the lack of absolute measurements.

5.4. By-products

The analysis of the site sometimes requires the extraction of complementary 2D products: line drawings, orthoimages and unrolled images.

5.4.1. *Line drawings*

In architectural studies of buildings, these are particularly common. They can be divided into several types (Figure 5.14):

– section: the object is intersected by a specified plane. For non-plane surfaces, a representation via parallel sections or contour lines can be used to take these geometric shapes into account. Equidistance is defined by the relief of the shapes, the scale of the survey and the required accuracy;

– cross-section: this is a section that also has the projection on the plane of the architectural lines of one part of the building. The topographic map is a special case of a horizontal cross-section;

– elevation: orthogonal representation of a façade on a plane.

The choice of a section plane can be made according to the analysis to be carried out. In the past, axonometric or isometric views also allowed the user to obtain representations from different angles. Nowadays, these methods are largely obsolete, having been replaced by direct 3D displays.

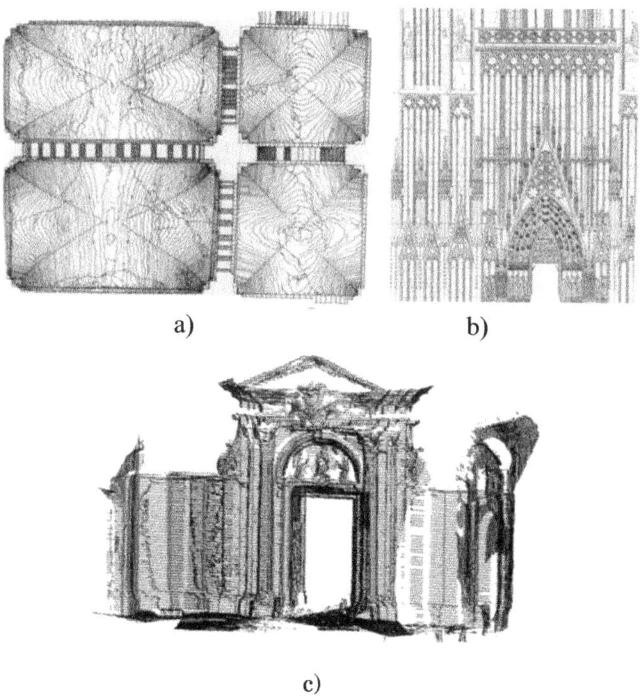

Figure 5.14. *a) Contour lines, b) elevations and c) parallel sections*

These products can also be created by a number of different methods:

– by a manual survey of the site, possibly supplementing it with several measurements from a distancemeter;

– by a survey with total station, also carried out in the field;

– by photogrammetric stereoplotting (see section 2.6);

– by plotting on point clouds: very often, software enables the isolation of slices on which the operator can plot the characteristic elements of interest (see section 5.3.2);

– by automatic processing of point cloud slices, automatically connecting points by polylines;

– by automatic processing of meshes: the calculation of the intersection of the model using a plane is carried out directly. Each face produces a line and when combined together, they make up a polyline. The disadvantage of this type of method is that the cross-section obtained in this way may be noised or inaccurate, depending on the quality of the reconstruction. In this case, it may be necessary to clean up the final result. There are also algorithms for increasing the regularity of polylines created and to reduce the weight of the products obtained [VAN 09];

– by direct plotting on a rectified image or an orthoimage (see section 5.4.2).

Once again, the main advantage of these products is that they are easily readable for all and they form the kind of representations that specialists are used to reading and analyzing.

5.4.2. *Production of orthoimages*

5.4.2.1. *Definition and use of orthoimages*

An orthoimage is an image that is projected orthogonally onto a plane (Figure 5.15). The plane is defined by the project sponsor, according to the requirements of the analysis. The orthoimage is georeferenced so that all the pixels are known in the global coordinate reference system. It is an image or a mosaic of images where the deformations due to the relief, the tilt of the viewing angle and the optical distortion are corrected.

An orthoimage is a metric image that can then be used as a basis for measurements. It is characterized by its extent, its spatial resolution (or ground sample distance: equivalent size of a pixel on the object), its spectral resolution (color, black and white, infrared, etc.), its accuracy, its coordinate system and the date of the image acquisition.

This product, which preserves the entire image's radiometric information, is therefore particularly rich in information. As with line drawings, being in 2D makes it immediately readable for any image processing software. The orthoimage sometimes requires supplementary operations for interpretation and plotting, especially to extract the metric information that it contains.

5.4.2.2. *The principle of orthoimage production*

Several images are usually necessary to cover the whole of the area, which needs to be processed. It is certainly possible here to reuse the photogrammetric images produced. In practice, the constraint of stereoscopic overlapping is not mandatory at this stage and more attention should be focused on the quality of the images, especially so that the resolution of the object is as consistent as possible, the angle of incidence is orthogonal and the images are radiometrically homogeneous.

As well as the images themselves, it is necessary to have a georeferencing of the images and a description of the shape of the object to be processed, in order to carry out geometric corrections of the orthogonal projection and the relief of the scene.

The orthorectification of each image is in itself automatic. It has two variants depending on whether it is a continuous surface being processed or a piecewise continuous surface, which is generally the case for outstanding sites and involves the presence of hidden parts. In the second case, one method consists of building a Z-buffer of the image, which specifies for each pixel the distance of a given point from the projection plane. This Z-buffer will show, for a given pixel, whether there is a corresponding one in the image or whether it is hidden [CHA 13]. In a pixel of the final orthoimage, one or several images can provide the information. The choice can be made by defining a criterion which, once again, may be based on resolution, incidence, continuity in the choice of images, a combination of these parameters, or even an average of the radiometric information available [GRA 07]. Some software perform an intelligent optimization of seamlines detection to minimize the visual transition from one orthophoto to the adjacent orthophoto.

Despite precautions taken during the photographic survey, it is often necessary to carry out a radiometric equalization of the images. This is because lighting conditions change very often from one image to another, as well as image surveying parameters and sometimes even the photographic equipment itself. Processing is therefore necessary to mask the visual differences between the images: these may be cosmetic corrections carried out manually with traditional image correction software or automatically as with a lot of photo stitching software now available, see, for example, [DAN 07].

Figure 5.15. *Orthoimage and elevation obtained by stereoplotting*

5.4.2.3. *Qualification of an orthoimage*

Other than the orthoimage parameters, which are the projection plane and the resolution, which are easy to verify using production metadata, the geometric quality of an orthoimage is mainly defined by its geometric accuracy. This depends on the quality of the 3D model, the georeferencing of the images, the production process and the mosaic method used. For each pixel of the orthoimage, geometric inaccuracies increase when the position of the image gets further away from the perpendicular of the projection plane, which passes through the point in question. Relative registration problems will be more visible at the edges of the object.

Geometric accuracy can be verified by using check points identified within the image and whose position in the reference system has been established by other methods. Measuring the distance between the two types of position can help to calculate the error average, which must be in accordance with the building specifications. It is also possible to compare the distances measured on the object and in the orthoimage, and also to verify the geometric continuity of the connections between the images. This assessment can be carried out very quickly, even if the information obtained in this way does not fully qualify the geometry.

Finally, radiometric quality takes into account the faithfulness of the image to the appearance of the scene in question and its ability to reproduce small intensity and color variations. In the absence of suitable measuring equipment, it is often estimated visually on a global level, and then more locally on the level of the seamlines between the images. It should be noted that the geometric inaccuracies may create blurring on the orthoimages in the case where the radiometry of each pixel is calculated according to a combination of images available.

5.4.3. *Other image products*

The orthogonal projection is well suited to façades and more general to sets of planes. In other cases, to limit deformations caused by projection or simply to make the object of interest readable, other types of transformations may be used: cylindrical projection in the case of a column [CHA 09], spherical projection for a dome, or a geodetic projection in more general cases (Figure 5.16).

a) b)

Figure 5.16. *Dome at the Al-Ashrafiya Mosque in Taiz (Yemen): a) orthogonal projection and b) a polar azimuthal equidistant projection*

5.5. Bibliography

[AIT 13] AITELKADI K., TAHIRI D., SIMONETTO E., *et al.*, "Segmentation of heritage building by means of geometric and radiometric components from terrestrial laser scanning", *Proceedings of the 24th CIPA Symposium*, Strasbourg, France, 2013.

[ALS 05] ALSHAWABKEH Y., HAALA N., "Automatic multi-image photo texturing of complex 3D scenes", *Proceedings of the 20th CIPA Symposium*, Torino, Italy, 2005.

[AME 01] AMENTA N., CHOI S., KOLLURI R., "The power crust", *Proceedings of the 6th ACM symposium on Solid modeling and applications, SMA'01*, ACM, New York, NY, pp. 249–266, 2001.

[BAL 13] BALLETTI C., BERTO M., GOTTARDI C., *et al.*, "Ancient structures and new technologies: survey and digital representation of the wooden dome of SS. Giovanni E Paolo in Venice", *Proceedings of the 24th CIPA Symposium*, Strasbourg, France, 2013.

[BAR 13] BARNEA S., FILIN S., "Segmentation of terrestrial laser scanning data using geometry and image information", *ISPRS Journal of Photogrammetry and Remote Sensing*, vol. 76, pp. 33–48, February 2013.

[BEC 09] BECKER S., "Generation and application of rules for quality dependent façade reconstruction", *ISPRS Journal of Photogrammetry and Remote Sensing*, vol. 64, no. 6, pp. 640–653, November 2009.

[BER 99] BERNARDINI F., MITTLEMAN J., RUSHMEIER H., *et al.*, "The ball-pivoting algorithm for surface reconstruction", *IEEE Transactions on Visualization and Computer Graphics*, vol. 5, no. 4, pp. 349–359, October–December 1999.

[BOR 11] BORRMANN D., ELSEBERG J., LINGEMANN K., *et al.*, "The 3D hough transform for plane detection in point clouds: a review and a new accumulator design", *3D Research*, vol. 2, no. 2, pp. 1–13, 2011.

[BOT 10] BOTSCH M., KOBBELT L., PAULY M., *et al.*, *Polygon Mesh Processing*, AK Peters/CRC Press, 2010.

[BOU 07] BOULAASSAL H., LANDES T., GRUSSENMEYER P., *et al.*, "Automatic segmentation of building facades using terrestrial laser data", *Proceedings of the ISPRS Workshop on Laser Scanning 2007 and SilviLaser 2007*, vol. 36, Espoo, Finland, 2007.

[BOU 10] BOULAASSAL H., CHEVRIER C., LANDES T., "From laser data to parametric models: towards an automatic method for building façade modelling", *Proceedings of the Third international conference on Digital heritage, EuroMed'10*, Springer-Verlag Berlin Heidelberg, pp. 42–55, 2010.

[BRY 13] BRYAN P.G., ABBOTT M., DODSON A.J., "Revealing the secrets of Stonehenge through the application of laser scanning, photogrammetry and visualisation techniques", *Proceedings of the 24th CIPA Symposium*, Strasbourg, France, 2013.

[CAR 01] CARR J.C., BEATSON R.K., CHERRIE J.B., *et al.*, "Reconstruction and representation of 3d objects with radial basis functions", *Proceedings of the 28th annual conference on Computer graphics and interactive techniques, SIGGRAPH'01*, ACM, New York, NY, pp. 67–76, 2001.

[CHA 09] CHANDELIER L., CHAZALY B., EGELS Y., *et al.*, "Numérisation 3D et déroulé photographique des 134 colonnes de la Grande Salle Hypostyle de Karnak", *Revue XYZ*, vol. 120, pp. 33–39, 2009.

[CHA 13] CHANDELIER L., EGELS Y., HÉNO R., *et al.*, "Photogrammétrie rapprochée pour l'architecture", Cours de l'Ecole Nationale des Sciences Géographiques, France, August 2013.

[CHE 09] CHEVRIER C., MAILLARD Y., PERRIN J.-P., "A method for the 3d modelling of historic monuments: the case of a gothic abbey", *Proceedings of the 3rd ISPRS International Workshop 3D Arch'09: "3D Virtual Reconstruction and Visualization of Complex Architectures"*, Trento, Italy, 2009.

[CHE 95] CHENG, Y., "Mean shift, mode seeking, and clustering", *IEEE Transactions on Pattern Analysis and Machine Intelligence*, vol. 17, no. 8, pp. 790–799, August 1995.

[CRO 13] CROMBEZ N., CARON G., MOUADDIB E.-M., "Colorisation photo-réaliste de nuages de points 3D", *Orasis, Congrès des jeunes chercheurs en vision par ordinateur*, Cluny, France, 2013.

[CUR 96] CURLESS B., LEVOY, M., "A volumetric method for building complex models from range images", *Proceedings of the 23rd Annual Conference on Computer Graphics and Interactive Techniques, SIGGRAPH'96*, ACM, New York, NY, pp. 303–312, 1996.

[DAN 07] D'ANGELO P., "Radiometric alignment and vignetting calibration", *Proceedings of the 5th International Conference on Computer Vision Systems, Camera Calibration Methods for Computer Vision Systems (CCMVS2007)*, Bielefeld, Germany, March 2007.

[DEM 11] DEMANTKÉ J., MALLET C., DAVID N., *et al.*, "Dimensionality based scale selection in 3D lidar point cloud", *International Archives of Photogrammetry Remote Sensing and Spatial Information Sciences*, vol. 38, part 5/W12, pp. 97–102, August 2011.

[DEK 03] DEKEYSER F., GASPARD F., DE LUCA L., *et al.*, "Cultural heritage recording with laser scanning, computer vision and exploitation of architectural rules", *International Workshop on vision techniques for digital architectural and archaeological archives*, Ancona, Italy, 2003.

[DEL 34] DELAUNAY D., "Sur la sphère vide", *Izvestia Akademii Nauk SSSR*, Otdelenie Matematicheskikh i Estestvennykh Nauk, vol. 7, pp. 793–800, 1934.

[DEL 07] DE LUCA L., VÉRON P., FLORENZANO M., "A generic formalism for the semantic modeling and representation of architectural elements", *The Visual Computer*, vol. 23, no. 3, pp. 181–205, 2007.

[DES 10] DESCHAUD J.-E., GOULETTE F., "Point cloud non local denoising using local surface descriptor similarity", *International Archives of Photogrammetry, Remote Sensing and Spatial Information Sciences*, vol. 38, part A, Saint Mandé, France, 2010.

[DEV 04] DEVEAU M., PIERROT-DESEILLIGNY M., PAPARODITIS N., *et al.*, "Relative laser scanner and image pose, estimation from points and segments", *International Archives of Photogrammetry, Remote Sensing and Spatial Information Sciences*, vol. 35, part 3A, pp. 1130–1135, Istanbul, Turkey, July 2004.

[DEV 05] DEVEAU M, PAPARODITIS N., PIERROT-DESEILLIGNY M., *et al.*, "Strategy for the extraction of 3D architectural objects from laser and image data acquired from the same viewpoint", *Proceedings of the ISPRS Working Group V/4 Workshop, 3D-ARCH 2005, "3D Virtual Reconstruction & Visualization of Complex Architectures"*, International Archives of Photogrammetry, Remote Sensing and Spatial Information Sciences, vol. 36, part 5/W17, Mestre-Venice, Italy, 2005

[DEY 05] DEY T.K., LI G., SUN J., "Normal estimation for point clouds: a comparison study for a voronoi based method", *Proceedings of Eurographics/IEEE VGTC Symposium Point-Based Graphics*, vol. 0, pp. 39–46, 2005.

[DIG 12] DIGNE J., "Similarity based filtering of point clouds", *Proceedings of the International Workshop on Point Cloud Processing, CVPR2012*, pp. 73–79, 2012.

[EDE 94] EDELSBRUNNER H, MÜCKE E.P., "Three-dimensional alpha shapes", *ACM Transactions on Graphics*, vol. 13, no. 1, pp. 43–72, January 1994.

[ELS 13] ELSEBERG J., BORRMANN D., NÜCHTER A. "One billion points in the cloud – an octree for efficient processing of 3D laser scans", *ISPRS Journal of Photogrammetry and Remote Sensing,* vol. 76, pp. 76–88, 2013.

[FAB 07] FABRIS M., ACHILLI V., BRAGAGNOLO D., *et al.*, "Filling lacunas in terrestrial laser scanning data: the "Cavallo ligneo" of the "Palazzo della Ragione" (Padua, Italy)", *Proceedings of the 21th International Symposium CIPA*, Athens, Greece, pp. 301–306, 2007.

[FIS 81] FISCHLER M.A., BOLLES R.C., "Random sample consensus: a paradigm for model fitting with application to image analysis and automated cartography", *Communication Association and Computing Machine*, vol. 24, no. 6, pp. 381–395, 1981.

[GAR 09] GARDON E., Fusion de cartes de profondeur en photogrammétrie architecturale, Rapport de stage du Mastère Photogrammétrie, Positionnement et Mesure de Déformation, National School of Geographical Sciences, France, 2009.

[GOD 09] GODET A., Création de modèle 3D pour le WEB du patrimoine Yéménite, Rapport de stage du cycle ingénieur 2ème année, National School of Geographical Sciences, France, 2009.

[GOS 13] GOSWAMI P., EROL F., MUKHI R., *et al.*, "An efficient multi-resolution framework for high quality interactive rendering of massive point clouds using multi-way kd-Trees", *The Visual Computer*, vol. 29, no. 1, pp. 69–83, 2013.

[GRA 07] GRAMMATIKOPOULOS L., KALISPERAKIS I., KARRAS G., *et al.*, "Automatic multi-view texture mapping of 3d surface projections", *Proceedings of the 3rd ISPRS International Workshop 2007: "3D Virtual Reconstruction & Visualization of Complex Architectures", 3D Arch'07, International Archives of Photogrammetry, Remote Sensing and Spatial Information Sciences*, vol. 36, part 5/W47, Zurich, Switzerland, 2007.

[HAM 11] HAMMOUDI K., "Contributions to the 3D city modeling: 3D polyhedral building model reconstruction from aerial images and 3D facade modeling from terrestrial 3D point cloud and images", PhD Thesis, University Paris-Est, Saint-Mandé, France, 2011.

[HAP 09] HAPPA J., MUDGE M., DEBATTISTA K., *et al.*, "Illuminating the past – state of the art", Proceedings of the 10th International Symposium on Virtual Reality, Archaeology and Cultural Heritage VAST – State of the Art Reports, 2009.

[HER 09] HERNÁNDEZ J., MARCOTEGUI B., "Point cloud segmentation towards urban ground modeling", *Proceedings of 5th GRSS/ISPRS Joint Workshop on Remote Sensing and Data Fusion over Urban Areas*, Shangai, China, 2009.

[KAZ 06] KAZHDAN M., BOLITHO M., HOPPE H., "Poisson surface reconstruction", *Proceedings of the Symposium on Geometry Processing*, pp. 61–70, 2006.

[KER 12] KERSTEN T., STALLMANN D., "Automatic tecture mapping of architectural and archaeological 3D Models", *International Archives of the Photogrammetry, Remote Sensing and Spatial Information Sciences*, Melbourne, Australia, vol. 34, part B5, pp. 273–278, 2012.

[KLA 09] KLASING K., ALTHOFF D., WOLLHERR D., *et al.*, "Comparison of surface normal estimation methods for range sensing applications", *Proceedings of the IEEE International Conference on Robotics and Automat ion, ICRA'09*, pp. 3206–3211, 2009.

[LAF 13] LAFARGE F., ALLIEZ, P., "Surface reconstruction through point set structuring", *Computer Graphics Forum*, vol. 32, no. 2, pp. 225–234, 2013.

[LAN 12] LANDES T., BOULAASSAL H., GRUSSENMEYER P., "Quality assessment of geometric façade models reconstructed from TLS data", *The Photogrammetric Record*, vol. 27, no. 138, pp. 137–154, June 2012.

[LEV 98] LEVIN D., "The approximation power of moving least squares", *Mathematics of Computation*, vol. 67, pp. 1517–1531, 1998.

[LIU 05] LIU Y., SNOEYINK J., "A comparison of five implementations of 3D delaunay tessellation", *Combinatorial and Computational Geometry*, vol. 52, pp. 439–458, 2005.

[LOR 87] LORENSEN W.E., CLINE H.E., "Marching cubes: a high resolution 3d surface construction algorithm", *Proceedings of the 14th annual conference on Computer graphics and interactive techniques, SIGGRAPH'87*, ACM New York, NY, pp. 163–169, 1987.

[MAI 08] MAÎTRE H. *Image Processing*, ISTE, London, John Wiley & Sons, New York, 2008.

[MAL 08] MALGOUYRES R., *Algorithmes pour la synthèse d'images et l'animation 3D*, Dunod, 2nd edition, Sciences SUP 2008.

[MEI 10] MEIERHOLD N., SPEHR M., SCHILLING A., *et al.*, "Automatic feature matching between digital images and 2d representations of a 3d laser scanner point cloud", *Proceedings of the ISPRS Commission V Mid-Term Symposium on Close Range Image Measurement Techniques*, vol. 38, Newcastle upon Tyne, United Kingdom, pp. 446–451, 2010.

[MUR 13] MURPHY M., MCGOVERN E., PAVIA S., "Historic building information modeling – adding intelligence to laser and image based surveys of European classical architecture", *ISPRS Journal of Photogrammetry and Remote Sensing,* vol. 76, pp. 89–102, February 2013.

[PER 10] PÈRE C., LANDRIEU J., ROLLIER-HANSELMANN J., *et al.*, "Reconstitution virtuelle de l'église abbatiale Cluny III: des fouilles archéologiques aux algorithmes de l'imagerie", *Actes du colloque Virtual Retrospect 2009, Pessac, 18–20 November 2009*, Archeovision, Ausonius Editions, Bordeaux, vol. 4, pp. 151–159, 2010.

[PFI 00] PFISTER H., ZWICKER M., VAN BAAR J., *et al.*, "Surfels: surface elements as eendering primitives", *Proceedings of the 27th annual conference on Computer graphics and interactive techniques, SIGGRAPH'00*, ACM New York, NY, pp. 335–342, 2000.

[PIE 13] PIERROT-DESEILLIGNY M., "MicMac, Apero, Pastis and Other Beverages in a Nutshell!", *Documentation de la suite Micmac*, 2013.

[PU 09] PU S., VOSSELMAN G., "Building Facade reconstruction by Fusing Terrestrial laser points and images", *Sensors*, vol. 9, no. 6, pp. 4525–4542, 2009.

[PUT 13] PUTTONEN E., LEHTOMÄKI M., KAARTINEN H., *et al.*, "Improved sampling for terrestrial and mobile laser scanner point cloud data", *Remote Sensing*, vol. 5, no. 4, pp. 1754–1773, 2013.

[RAB 06] RABBANI T., HEUVEL van dEN F.A., VOSSELMAN, G., "Segmentation of point clouds using smoothness constraints", *International Archives of Photogrammetry, Remote Sensing and Spatial Information Sciences*, Dresden, Germany, vol. 36, part 5, pp. 248–253, 2006.

[SCH 07] SCHNABEL R., WAHL R., KLEIN R., "Efficient RANSAC for point-cloud shape detection", *Computer Graphics Forum*, vol. 26, no. 2, pp. 214–226, 2007.

[SMI 11] SMIGIEL E., ALBY E., GRUSSENMEYER P., "TLS data denoising by range image processing", *Photogrammetric record*, vol. 26, no. 134, pp. 171–189, June 2011.

[SMI 13] SMIGIEL E., ALBY E., GRUSSENMEYER P., "Terrestrial laser scanner data denoising by dictionary learning of sparse coding", *Proceedings of the 24th CIPA Symposium*, Strasbourg, France, 2013.

[STE 04] STERNBERG H., HERSTEN T., JAHN L., *et al.*, "Terrestrial 3D laser scanning – data acquisition and object modeling for industrial as-built documentation and architectural applications", *International Archives of the Photogrammetry, Remote Sensing and Spatial Information Sciences*, Istanbul, Turkey, vol. 35, part B7, pp. 942–947, 2004.

[VAN 09] VAN GENECHTEN B., SANTANA QUINTERO M., DEMEYERE T., *et al.*, "Terrestrial laser scanning in architectural heritage – deformation analysis and the automatic generation of 2D cross-section", *Proceedings of the CIPA 22nd International Symposium*, Kyoto, Japan, 2009.

[VOS 04] VOSSELMAN G., GORTE B.G.H., SITHOLE G., *et al.*, "Recognising structure in laser scanner point clouds", *International Archives of Photogrammetry, Remote Sensing and Spatial Information Sciences*, Freiburg, Germany, vol. 46, part 8/W2, pp. 33–38, October 4–6, 2004.

[WAN 04] WANG M., TSENG Y.-H., "LIDAR data segmentation and classification based on octree structure", *International Archives of the Photogrammetry, Remote Sensing and Spatial Information Sciences*, vol. 20, part B3, Istanbul, Turkey, 2004.

[ZHA 09] ZHAN Q., LIANG Y., XIAO Y., "Color-based segmentation of point clouds", *Proceedings of ISPRS Laserscanning09 workshop*, Paris, France, vol. 38, part 3/W8, pp. 248–252, 2009.

6

Management and Use of Surveys

6.1. Introduction

The creation of a 3D model of an outstanding site as described in the previous chapters is only useful if the requirements which initially motivated it can be comfortably satisfied by the users at any point. This means that the data must be available in the correct format, in the right place and at the right time. Indeed, the current surveying methods for outstanding buildings produce varied data (in nature and format), which obviously have large volumes (large quantity of images, point clouds with several million points). This becomes even more significant since as the performance of acquisition equipment and software improves, the price of digital storage gets even lower. Following the initial excitement caused by new surveying techniques (lasergrammetry and more recently "fully automatic" photogrammetry), or the possibility of participative digitization, key questions need to be answered regarding the conservation and broadcasting of these data.

6.2. Managing data conservation

The surveying methods discussed in this text can be used to produce high-quality data, by combining cutting-edge technology with procedures developed by experts experienced in the production of highly accurate metric data. It is reasonable to make the price of these data cost-effective by making it available to be shared and evolved upon. To do so, we must ensure that it is:

– consistent: data must have been purged of any intermediary versions. Moreover, with the use of quality metadata, we should be able to assess the value of archived data;

– visible: since knowing that a digital survey already exists means that you can avoid re-doing existing work;

– readable: existing surveys must be able to be consulted by others, beyond the group of experts who participated in their production;

– accessible by parts: it is important to be able to consult a specific part of the outstanding building;

– durable: the storage facilities and file formats used must be durable.

6.2.1. *Backup methods*

For lack of common standard, judgment must be made on what should be recorded:

– all raw output data from the instrument: images in raw format, laser point clouds and raw topometric observations;

– all processed data: catalogue of ground control points, images with georeferencing, photogrammetric point clouds and photogrammetric stereoplotting;

– all deliverable data: processed point clouds, complete "packaged" 3D model and report.

[HUC 11] advises eliminating all temporary documents and provisional versions beforehand. All data (apart from raw data) must be renamed so that it is more comprehensible. Several copies are made on recording media, which are suitable for the volume of data, one of which should be conserved remotely.

It is not mandatory to conserve intermediary data (tie points between images, image pyramids, etc.), but it should be possible to regenerate them, especially in the case where work needs to be continued or extended on the site. This is why it is essential to rigorously document the stages of production (name and version of software used, commands, parameters and qualification of intermediate results).

6.2.2. *Semantic enhancement of geometric data*

Surveys of outstanding buildings as they are discussed here are essentially composed of geometric elements: points, lines, surfaces or meshes which are occasionally textured by the images. These elements need to be enhanced with the use of semantic information. This represents a real element of added-value for users, and also makes accessibility of the data easier. However, the way in which the surveyed data are described (date of survey, team responsible, methods used, quality obtained or nature of the building or parts of it, etc.) can obviously differ: what may be clear in one case, may be unclear in another. The metadata must, therefore, be standardized, or derived from ontology used in the surveying field [VRO 08].

Beyond the essential reflection on the form of the semantic metadata, the way in which it is carried out is also important, given the volume of the data involved. Some information is recorded during the acquisition (e.g. exchangeable image file format (EXIF) data for images) and the processing (calculation reports), but the majority must be taken down in an interactive way, by filling in a form, for example. Some automations can be expected: [MAN 13] suggests automatically communicating the annotations made manually on raw images to all overlapping images through the 3D point cloud. Search techniques for images and parts of images by visual content can also be used to label objects in images. Traditionally, the techniques consist of two stages: extraction of characteristics, such as color and shape, then resemblance measurements. In the architectural field, the semantic description of a surveyed element may be based on catalogs of shapes adapted from architectural pattern books [DEL 11].

6.2.3. *Standardization*

The concern that data acquired today will not be re-usable outside the structure where they have been produced, or will be obsolete several years later, gives rise to the need for standards and rules in the field of surveying outstanding sites.

Admittedly, the various types of data acquired during a survey are naturally recorded in either standard or very widespread formats: TIFF[1], JPEG, JPEG 2000[2] for images, LAS[3] mainly for aerial point clouds, STL[4] and PLY[5] for

1 Tagged Image File Format, http://partners.adobe.com/public/developer/tiff/index.html.

2 http://www.jpeg.org/jpeg2000/index.html.

3 Exchange format of laser data, http://www.asprs.org/a/society/committees/standards/LAS_1_4_r13.pdf.

4 http://en.wikipedia.org/wiki/STL_%28file_format%29.

5 Polygon File Format, http://paulbourke.net/dataformats/ply/, format may also be used for the storage of point clouds.

meshes as well as point clouds, OBJ[6] which is also compatible with other types of 3D models, and PDF3D[7] for the publishing of data. There are also standard formats for derivative products (vector format 3D models). The CityGML[8] format, created for modeling urban environments, is worth examining. Indeed, it allows a town to be modeled, down to the individual building, including georeferenced 3D vector geometry, 3D topology (the link between the different entities in the model), semantics and texture. This format, standardized by the Open Geospatial Consortium (OGC) in 2008, offers five LOD's (levels of detail), from LOD0 to LOD4, where the fourth and fifth, LOD3 and LOD4 are suited to data produced for the survey of an outstanding site (description of an architectural building, including the interior and exterior). The IFC[9] is also an object-based file format used by the construction industry to exchange and share information between different software, and can be useful for storing models of buildings, although it processes texturing and georeferencing in a more basic way than CityGML. These standards (CityGML and IFC) only concern the post-processed part of the data, because raw point clouds, for example, which are abound with fine detail on the surveyed building, are not dealt with. Here, other 3D standards will not be discussed (X3D[10], COLLADA[11]), as they are mainly designed for visualization applications.

Furthermore, the survey of an outstanding site is much more than a vector model or a collection of sparsed images and point clouds: it is an ordered and structured collection, which must be conserved as such, allowing others to access

6 http://people.sc.fsu.edu/~jburkardt/txt/obj_format.txt.

7 http://www.pdf3d.fr.

8 http://www.citygml.org.

9 http://www.ifcwiki.org/index.php/Main_Page.

10 www.web3d.org/x3d.

11 https://collada.org.

and interpret them. The advantages and disadvantages of using a standard in a service carrying out surveys are given in [MID 12]. Unsurprisingly, complying with a standard can make the general production process more onerous, but increases the credibility of the data produced, and makes it easier to re-use at a later date.

The field of documenting historical architectural heritage, where very high-resolution data – either raw images or 3D point clouds – is frequently used for analysis, has been extremely active for several years on questions of the sharing and interoperability of surveys. The CIPA[12], one of the ICOMOS committees (International Council for Monuments and Sites) specialized in documenting cultural heritage, in partnership with the ISPRS (International Society for Photogrammetry and Remote Sensing), which campaigns for durable and shareable data, is promoting the application of protocols (London Charter[13], EPOCH[14], Seville principles [LOP 11]), and moreover encouraging the creation of suitable standards for historical architectural heritage [IOA 05].

There are already open technologies and standard protocols for the interoperability of national and European metadata (Dublin Core[15], EAD[16], OAI-PMH protocol[17]), and free software cataloguing tools (e.g. Geonetwork[18]), but they do not satisfy all the requirements regarding the management of 3D geometric data for the survey of outstanding sites.

12 International Committee for Documentation of Cultural Heritage.

13 http://www.londoncharter.org/.

14 http://www.epochnet.org/index.php?option=com_content&task=view&id=212&Itemid=357.

15 http://dublincore.org/.

16 http://www.loc.gov/ead/.

17 http://www.openarchives.org/pmh/.

18 http://geonetwork-opensource.org/.

6.2.4. *Responsibility and support for data conservation*

The UNESCO charter for the preservation of digital heritage[19] recommends that digital heritage be managed by a legal institution, and that the notion of a "legal deposit" be extended into the field of heritage. This recommendation would then make it possible to see whether a geometric survey has already been carried out on a given site or not, but it seems to be at odds with the economic and competition issues in the private sector. [IOA 05] points out that in his experience, only public bodies are able to promote the use of standards, and more generally, good practices regarding interoperability of surveys of outstanding sites. Europeana[20], the digital library launched by the European Commission in 2008, aims to include surveys of outstanding sites amongst other objects within its museum collection. Member states also have databases, which describe the sites (e.g. *atlas des patrimoines*[21] in France), which include, when they exist and have been indicated, a link to 3D digital surveys.

6.3. Diffusion of data

6.3.1. *Expected functionalities*

The data to be made available (final 3D model and any intermediary data, such as georeferenced images or point clouds, ground control points, etc.) should be accessible through a catalogue, through local access or on the internet:

– by geographical extent: overall relationship between volume parameters, for example, and an overall plan. Therefore, for a cathedral, this could be the loading of point clouds and of the vector model intersecting the cube of an

---

19 http://portal.unesco.org/en/ev.php-URL_ID=17721&URL_DO=DO_TOPIC&URL_SECTION=201.html.

20 http://www.europeana.eu/.

21 http://atlas.patrimoines.culture.fr.

edge of 5 m centered on the point of intersection of the nave and the transept;

– according to the nature of the entities: e.g. loading point clouds which describe the main exterior doorway;

– by semantic requests: e.g. loading data acquired before 1995;

– by viewpoint defined by digitization stations: e.g. loading of everything that can be seen from a given viewpoint;

– or by any viewpoint: interpolation between views from digitization stations.

These functionalities must be able to be combined (e.g. on a given structure, loading of orthoimages covering the Southern triforium acquired before 2005). Finally, access must be multi-resolution, and allows the loading not only of reconstructed models, which are necessarily a little simplified, but also of 3D points if they exist, at their full resolution.

Management strategies for high-volume data discussed in section 5.1 are clearly essential for the easy accessibility of data. The representation of elements of the building according to various LODs can also improve the ease of navigation within the data by showing the level of detail adapted to the distance from the viewpoint, and therefore avoiding the necessity of loading all the data.

6.3.2. *GIS*

The above requested functionalities are clearly similar to those of a geographic information system. Current software, some of which can be processed in 3D[22], are above all suitable for representations of districts or towns. They are not yet fully compatible for the management of survey data from an outstanding site. They are unable to process large point clouds in particular, and to represent complex shapes, which make up the monument. However, it is natural to organize data in a spatial database.

6.3.3. *Web platforms*

Virtual globes allow the visualization of correctly formatted data of reasonable dimensions (not 3D point clouds but vectorized models). Thus, the KML[23] format, a file format based on XML coding, enables the modeling and storage of characteristics, such as points, lines, images, polygons and models, for displaying on Google Earth.

There are also other platforms, which are more specifically dedicated to the visualization and analysis of surveys of outstanding sites, including the following:

– cyark[24]: a world heritage documentation platform, works as a kind of aggregator for cultural site surveys. The sites are described by different media: texts, archive drawings, videos and occasionally 3D models (3D point clouds mainly from lasergrammetry). For some sites discussed, it is possible to visualize and manipulate the sub-sampled laser data, which are already divided into separate zones (e.g. Beauvais cathedral[25] in an integrated 3D visualizer);

22 http://resources.arcgis.com/en/help/main/10.1/#/What_is_the_ArcGIS_3D_Analyst_extension/00q8000000wv000000/.

23 Keyhole Markup Language, https://developers.google.com/kml/documentation/.

24 http://archive.cyark.org/.

25 http://archive.cyark.org/cathedral-of-beauvais-gallery.

– mappinggothic.org[26]: founded by two American professors of art history, this site brings together a large variety of documents on many Gothic sites in Europe (mainly France and UK). For each site, text information is available (history of the construction, historical context, etc.) as well as panoramic views, anaglyph stereo-pairs, very high-resolution images of some decorative elements and images of laser point clouds. Layouts of compared characteristic elements of the various cathedrals modeled are also available;

– 3D monuments[27]: program for visualizing and evaluating 3D models of national monuments, via the Nubes platform, developed and maintained by the MAP (*Modèles et simulations pour l'Architecture et le Patrimoine* – UMR 3495). This platform allows the user to see the data acquired and the processing carried out on certain sites, by the way of images and videos.

Evidently, this list is not exhaustive, but it appears that no web platform is yet available, which enables the complete dynamic manipulation of all its data produced during a survey. Nevertheless, it is important to note [DEV 12] who presents a high-performance web platform for visualization and interaction with images and point clouds, using the webGL[28] technology, and [GUA 09] whose light web client application enables the visualization and editing of 3D models. In the field of lasergrammetry surveys, some manufacturers provide simple applications for visualizing and sharing data, including several manipulation tools. This software can prove to be sufficient for some measurement and analysis operations[29].

26 http://mappinggothic.org/.

27 http://www.map.archi.fr/3D-monuments/.

28 http://www.khronos.org/webgl/.

29 http://hds.leica-geosystems.com/en/Leica-TruView_63960.htm.

6.4. Bibliography

[DEL 11] DE LUCA L., BUSAYARAT C., STEFANI CH., *et al.*, "A semantic-based platform for the digital analysis of architectural heritage", *Computers & Graphics,* vol. 35, no. 2, pp. 227–241, April 2011.

[DEV 12] DEVAUX A., PAPARODITIS N., BRÉDIF M., "A web-based 3D mapping application using WebGL allowing interaction with images, point clouds and models", *20th ACM SIGSPATIAL International Conference on Advances in Geographic Information Systems (ACM SIGSPATIAL GIS 2012),* Redondo Beach, CA, USA, 6–9 November 2012.

[GUA 09] GUARNIERI A., PIROTTI F., VETTORE A., "An open source application for interactive exploration of cultural heritage 3D models on the web", *Proceedings of 3D-ARCH 2009: 3D Virtual Reconstruction and Visualization of Complex Architectures,* vol. 38, ISPRS Archives, part 5/W1, WG V/4, Trento, Italy, 25–28 February 2009.

[HUC 11] HUC C., "Le risque de tout perdre, Archivage numérique", *Géomètre,* vol. 2080, pp. 46–48, 2011.

[IOA 05] IOANNIDES M., GEORGOPOULOS A., SCHERER M., "Standards in cultural heritage: the missing grammar for the digital documentation of the past", *Proceedings of the 20th CIPA symposium*, Turin, Italy, 2005.

[LOP 11] LOPEZ-MENCHERO V.M., GRANDE A., "The principles of the Seville charter", *Proceedings of the 23rd CIPA symposium*, Prague, Czech Republic, 2011.

[MAN 13] MANUEL A., STEFANI C., DE LUCA L., *et al.*, "2D/3D semantic annotation towards a set of spatially-oriented photographs", *Proceedings of the 24th CIPA symposium*, Strasbourg, France, 2013.

[MID 12] MIDAS HERITAGE, The *UK Historic Environment Data Standard (V 1.1),* English Heritage, London, *2012.*

[VRO 08] VROCHIDIS S., DOULAVERAKIS C., GOUNARIS A., *et al.*, "A Hybrid Ontology and Visual-based Retrieval Model for Cultural Heritage Multimedia Collections", *International Journal of Metadata, Semantics and Ontologies,* vol. 3, no. 3, pp. 167–182, 2008.

Conclusion

It is unreasonable to talk of a conclusion when the methods discussed in this text for documenting outstanding sites are, as shown here, in a period of rapid evolution. The equipment, whether laser or photo, is becoming lighter, more effective, less expensive and easier to manipulate. The software is becoming more automatic and can be turned to in more and more complex situations. It can be used today on sites which would have until recently been thought of as impossible to work with. However, never-ending progress in the field of computer science means that this type of complexity can now be processed.

Nevertheless, the performance and the modernity of these methods have not changed the underlying values: in [SAI 92], we are reminded that the survey "must establish the shapes of the object, their proportions and dimensions, and clearly describe the multiple relations between the various elements". Saint-Aubin highlights the fact that "an observation of the representation is insufficient for establishing what is surveyed and what is interpolated and, even more seriously, what is accurate or inaccurate in the elements surveyed". This call for caution, given in 1992 at the dawn of the first laser scanners and well before the automation of photogrammetry, is highly relevant today. Data produced using the current methods, now available to those outside the selected group of surveyors, are only useful if accompanied by metadata which qualifies it in a methodical way, and describe the fabrication process. It is important to insist once again, as in Chapter 6, that the accessibility of these data is essential for optimizing its effectiveness. Progress in this field is eagerly awaited, especially given that at the same time, surveying methods continue to evolve, making data production ever-simpler.

Apart from these reservations, it is important to reiterate the interest of modern lasergrammety and photogrammetry for analyzing and documenting the outstanding sites. Their enormous contribution to heritage conservation cannot be overstated. Recent examples of sites, which have been destroyed (the Buddhas of Bamiyan, in Afghanistan, destroyed in 2001, and many UNESCO world heritage sites in Syria which have been seriously damaged by the civil war), remind us of the importance of having a documentation which is able, at least, to preserve their memory, but which could also enable faithful reconstructions.

C.1. Bibliography

[SAI 92] SAINT-AUBIN J.-P., *Le Relevé et la Représentation de l'architecture,* Documents et méthodes, vol. 2, 1992.

Index

Q, R

S

T

U, V, W, Z

Other titles from

iSTE

in

Geographical Information Systems

2013

SALLABERRY Christian
Geographical Information Retrieval in Textual Corpora:

2012

BUCHER Bénédicte, LE BER Florence
Innovative Software Development in GIS

2011

BANOS Arnaud, THÉVENIN Thomas
Geographical Information and Urban Transport Systems

DAUPHINÉ André
Fractal Geography

2010

BRUNET Roger
Sustainable Geography

CARREGA Pierre
Geographical Information and Climatology

CAUVIN Colette, ESCOBAR Francisco, SERRADJ Aziz
Thematic Cartography – 3-volume series
Thematic Cartography and Transformations – volume 1
Cartography and the Impact of the Quantitative Revolution – volume 2
New Approaches in Thematic Cartography – volume 3

LANGLOIS Patrice
Simulation of Complex Systems in GIS

MATHIS Philippe
Graphs and Networks – 2nd edition

THERIAULT Marius, DES ROSIERS François
Modeling Urban Dynamics

2009

ROCHE Stéphane, CARON Claude
Organizational Facets of GIS

2008

BRUGNOT Gérard
Spatial Management of Risks

FINKE Gerd
Operations Research and Networks

GUERMOND Yves
Modeling Process in Geography

KANEVSKI Michael
Advanced Mapping of Environmental Data

2007

DUMOLARD Pierre
Spatial Interpolation for Climate Data

SANDERS Lena
Models in Spatial Analysis

2006

CLIQUET Gérard
Geomarketing

DEVILLERS Rodolphe, JEANSOULIN Robert
Fundamentals of Spatial Data Quality

Printed and bound by CPI Group (UK) Ltd, Croydon, CR0 4YY
13/09/2026
14961081-0001